A PIMPERNEL GARDEN CLASSIC

THIS BOOK BELONGS TO

MEADOWS

AT GREAT DIXTER AND BEYOND

A Pimpernel Garden Classic

Meadows

at Great Dixter and Beyond

CHRISTOPHER LLOYD

with a new introduction by

FERGUS GARRETT

photographs by JONATHAN BUCKLEY
and CAROL CASSELDEN

Pimpernel
Press ltd
www.pimpernelpress.com

Pimpernel Press Ltd
www.pimpernelpress.com

First published in Great Britain as *Meadows* in 2004 by Cassell Illustrated

This revised edition published by Pimpernel Press Limited 2016

A catalogue record for this book is available from the British Library

Typeset in Adobe Caslon

ISBN 978-1-910258-03-3
Printed and bound in China

9 8 7 6 5 4 3 2 1

PAGE 1 Originating in camas prairies of North America, *Camassia quamash* has naturalized in pockets throughout the front meadow at Great Dixter, ably coping with dense competition and happily rubbing shoulders with native buttercups and grasses.
PAGE 2 Snakeshead fritillaries (*Fritillaria meleagris*) have self-sown throughout the orchard at Dixter, doing best in semi-shaded sites under trees but also happily stretching out into open spaces.
RIGHT Early morning mist over the narrow brick wheelbarrow path which cuts through the lower orchard, joining the Long Border to the nursery, highlighting the undulations of the ground and defining the long grass.

CONTENTS

6 INTRODUCTION
By Fergus Garrett

48 INTRODUCTION TO THE FIRST EDITION
By Christopher Lloyd

54 1. NATURE, AND THE HAND OF MAN

72 2. A GARDENER'S MEADOWS

84 3. A SETTING FOR WILDFLOWERS

90 4. CREATING PERMANENT MEADOWS

104 5. MEADOW MANAGEMENT

112 6. MAINSTAY GRASSES

118 7. PERENNIALS

156 8. BULBS & CORMS

180 9. THE CORNFIELD SCENE

192 10. PRAIRIES

214 11. THE WIDER LANDSCAPE

228 Bibliography
229 Index
240 Acknowledgements

Introduction

My first visit to Great Dixter was in 1985 as a horticultural student from Wye College. Christopher Lloyd (or Christo, as he was known to his friends) also studied there, immediately after the Second World War, and he then joined the teaching staff for a short period, before returning to Dixter to set up the Great Dixter Nurseries, write, and garden from his home. A natural teacher, Christo influenced many people through his articles, books and lectures. One of his early students was Tom Wright, who then went on to teach me during my time at Wye. Tom kept a close relationship between the college and Christo, and regular coach outings by students to Dixter became customary.

At nineteen years old I was pretty green and inexperienced, having had very little to do with gardens or gardeners. As a student, I'm ashamed to say, I hadn't heard of Great Dixter or Christopher Lloyd. I'm even more ashamed to say that I didn't read up on the place prior to our visit and took it as just another day out. Little did I know that this event would be life-changing.

FIRST IMPRESSIONS

The coach rolled through the lush countryside between Wye and Northiam (where Great Dixter is situated). The Wealden landscape moulded in a patchwork of fields and woods, segregated by hedging, was punctuated by solitary trees and occasionally interrupted with sprawling villages. As we pulled up, my first glimpse of the house was through the pendulous branches of ancient oak trees overhanging the private driveway to the front gate. The building was imposing: immense in size, but elegant and low-hung in its setting and comforted by trees and vegetation that hugged its sides. We gathered outside the entrance and made our way through the simple old oak gate down the rickety path of York stone, with long meadow grass either side of us. It must have been late June, as the meadows

On either side of the front path long grass welcomes you, framing the view of the house. The greeting party is often two tenacious dachshunds, which have been a hallmark of life here for many decades.

were mainly flowerless and brown, with just flower heads of the grasses adding texture. Christo was waiting for us, with the old leaning porch behind him tilting precariously to the east.

My first impression of him was curious – he looked a bit like Einstein, with a neatly trimmed moustache and a shock of white hair. He was dressed in a checked flannel shirt worn ragged at the collar and cotton drill trousers muddy at the knees, with holes in his shoes (he hated buying new shoes). With rosy cheeks and searing blue eyes, and hands cupped behind his back, he seemed happy to have us there although a little uneasy in new company.

Christo took us round as a group – taking ages – stopping at almost every plant. Notebook in hand and hungry to learn, I followed each footstep, hanging on every word. The gardens, wrapped round the house, were subdivided into a series of rooms by hedges and buildings. Each compartment was a different experience, yet all fitted together like the pieces of a complex jigsaw. Pockets of meadow interrupted the scene here and there, in parts coming deep into the garden and occasionally opening the views into the landscape beyond. A large expanse of long grass dominated the south side, flowing along the famous Long Border and separated from it merely by a narrow strip of mown grass and a long path of York stone. My only previous gardening experience had been with Brighton Parks, so the full Dixter encounter came as bit of a shock – not just for me, but for the others too; many of us had never seen a garden like it. Tom Wright knew full well that this place with its style and unconventionality would test us. After a long tour with Christo, we were given free time to roam.

On our return to college we were asked to write up our visit. I can't recall much about the report that followed – no doubt it would make fascinating reading for me today! However, what I do remember are my remarks on the meadows. Interestingly, my first impressions then can't have been too dissimilar from those of many people now. It seemed clear to me that the long grass within the garden was out of place: it looked messy and dragged the garden down. Having since become familiar with the meadows, I realize that this impression was due partly to the time of year, and partly to a lack of awareness on my part. Late June is never a good time to see the long grass at Dixter, with much of the colour gone, unless you are predominantly interested in the grasses themselves. The common spotted orchids, ox-eye daisies, buttercups and clover have finished, and the knapweed and later bird's-foot trefoil haven't yet started. It's a bit much to assume that a first-time visitor will see great beauty in seed heads, or embrace the colour brown and the shagginess of it all. And no doubt my reaction would have been different had I visited in April or May, when the meadows are alive and fresh, filled with an ever-changing tapestry of flowers, swaying in the wind and dripping with the rich beauty typically associated with the English countryside.

An atmospheric meadow scene at the front of the house, as early morning dew settles on the seed heads of fine grasses such as common bent (*Agrostis tenuis*) and lesser knapweed (*Centaurea nigra*), with added colour from the late-flowering tufted vetch (*Vicia cracca*) bringing sprinklings of purple and mauve among the froth of biscuit browns. All of this is framed by the dark green yew of Great Dixter's hedges.

In late summer the meadows take on the appearance of a turbulent sea, with fawn-coloured grasses brushed and blown apart, lodged and unsettled by rain and wind. This late stage has its own beauty coupled with an air of senescence as the seed is ripe and the meadows are ready to cut. Cutting is done slowly, taking one area at a time, allowing the animals to move and the seed to spill.

Thinking back to this moment in my life I'm not surprised at all that I thought that these 'messy' areas should be outside the garden boundary, and that they belonged in the countryside and not within an ornamental setting where people pay an entrance fee and expect to see something more manicured. Tom Wright questioned my views but I now know that I was too immature and my senses too underdeveloped to see it any other way. I reacted negatively to Dixter's wilderness having been accustomed to an environment in which total control of nature was the norm, with very little acceptance of anything else. Not much was said about wildlife at home or at school. Nature surrounded us but we grew up ignorant of it. Even when I was at university, horticulture was all about taming wilderness and, were it not for Tom Wright, our three years of studying could easily have passed us by without much consideration of the subject.

While at Wye I became obsessed by garden plants. My notebooks were filled with Latin names, with colourful descriptions and sketches of petals and flower shapes. Looking at my notes now it amazes me that whenever I visited a garden I hardly remarked on anything to do with atmosphere, or spirit or sense of place, or the landscape. I never for once took into account how a gem like Dixter would connect to its surroundings, let alone to natural life and biodiversity within and outside its boundary. At Wye, rural environmental scientists studied the environment and horticulturalists studied gardening – the crossover was limited. But it was that visit to Great Dixter and meeting Christo that paved the way to my being more aware of everything that surrounded me. As is often the case, this response didn't happen overnight – the thread that connected me with Mother Nature remained hidden for many years. I started late, and so teasing this sensitivity into the open was a gradual process, developing and accelerating as I matured.

Christo liked the way I took notes – we struck up a friendship and kept in touch by letter. I often visited him at weekends along with another Wye student, Neil Ross. Neil and I both loved being at Dixter, surrounded by amazing garden plants, vibrant colour and always new and exciting combinations. My fixation with individual plants continued, and I ticked them off one by one like an obsessive twitcher, encouraged by the stamp-collector mentality that sits so embedded in many young people – especially boys. Christo's encyclopaedic knowledge fuelled my fire and the lists in my notebooks grew and became more exotic with every visit.

But being with Christo wasn't just about generating lists. He made me look at everything – opening up my senses with long walks into the woods and walking across the fields at Dixter. He made me look at birds, butterflies, fungi, mosses, and made me aware of the changing seasons, the sounds and smells of the countryside, from autumn colour on the trees to leaves on the ground. He was close to nature and through him I became acutely observant and suddenly my surroundings took on another meaning.

APPRECIATING THE WILD

Shortly after Wye, a friend I had met through Christo accompanied me on a trip to Turkey, where I had spent a great deal of my childhood. We went to the Western

Pontic Alps around Bursa, venturing into the rolling hills of high beech forest, with pockets of bay and arbutus, sweet chestnut, hazel and scrub oak. The woodland floor had snowdrops and scillas in flower, with hellebores scattered here and there, growing with ferns, *Daphne pontica* and butcher's broom. Cistuses and broom habited the more open areas, with clematis scrambling through the scrub. Even though it was February and patches of snow covered the ground, everywhere you looked there was something. Seeing garden plants in their natural habitat was a revelation and suddenly the relationship between you and them changed. Entering their territory you were no longer in control: you were in another world and it was magic.

Even a winter's walk down to the beach on the shores of the Marmara Sea became a botanical outing. They were littered with halophytic plants, skeletons of sea holly, viper's bugloss and the tortured deciduous bones of various shrubs, with *Clematis viticella* scrambling up wind-battered roses. The wasteland around our house in Istanbul – where, as children, we had pitched our tent, lit camp fires and played – came to life. There were sweeps of the dainty grass *Oryzopsis miliacea*, dried in the winter wind, dancing elegantly at waist height; sharing the ground were thistle heads and stalks of hollyhocks and little drumstick alliums sticking out their dehisced heads like a lollipops. As children this was the place where we ran loose. Now with newly found vision I could imagine it coming alive with a beautiful tapestry of red poppies, mallows, magenta geraniums and metallic blue sea hollies.

My fascination with the wild grew deeper. Although my city upbringing hadn't laid much of a foundation for this, people and experiences paved the way, and it didn't take long for an impressionable young man to fall in love with the beauty of an untamed world. Now, every time I returned to Dixter I viewed the garden differently. I, of course, saw artistry in Christo's association of flowers; his garden was a madcap canvas where he let loose his imagination, and Dixter's borders impressed me from one visit to another. But I also saw beauty in the quieter, wilder areas. With new eyes, the meadows came to life – complex, never in equilibrium, yet soft, gentle and billowing and with their own dynamic adding another dimension to the garden. Now, even during the brown season, my intrigue with this community didn't weaken – a far cry from my first visit.

Immersing my mind in the meadows became an important part of every visit to Dixter. Like the garden they changed from one instant to another and from one year to the next. The complexity of the community, with one wave following another, fascinated me. The season started early: carpets of crocuses splayed fully open from the end of February, praying to the sun, their petals backlit with low early season light, opening and shutting with the fluctuating temperatures and visited by drunken bees. Crocuses soon gave way to sweeps of daffodils warming the land in their joyous yellow. Orchids, cow parsley, buttercups, meadow cranesbills, vetches, clover and knapweeds among grasses followed. One layer of vegetation integrated and overlapped with another, each and every plant jostling for room within a picturesque, mysterious and dynamic matrix.

OVERLEAF The orchard at Great Dixter with large sweeps of old-fashioned daffodils, above which tower the trunks of wild pears, crab apples, hawthorns, and fruit trees. The trees are pruned sympathetically, with their lower branches allowed to sweep down and embrace the meadow.

After Wye I worked in a number of gardens. By now I had developed a special interest in garden design. Initially, I was struck by slick hard landscaping framing the space, with symmetrical water features, supported by architectural and fashionable muscular planting. With time, I became more open to a softer picture, trying to evoke a different, calmer atmosphere and wherever possible forming a link with nature. Meadows played an important part in all this, providing a contrast to formality, resting the eye, giving comfort, tempering all the hard edges as well as changing the pace within these formal spaces. Whether occupying large areas or smaller segments, whether distinguished from the formal space by strips of neatly mown lawn or by path or hedge, whether dominating the whole picture or perhaps occupying a small corner, meadows offered breathing space and incomparable movement. They also offered a glimpse of times gone by – the regeneration of a lost habitat and a trip into the past. I loved using them but size was always an issue: they weren't so easy to integrate into small urban and suburban spaces and they needed defining with a frame around them. But they also needed explaining to the client; meadows may be in vogue now but they were far from it in the 1990s. Of all the elements in design they proved to be the most divisive, often met by the client with silence as you explained away.

Then there is another aspect to consider. Meadows to me are as much about wildlife as flowers. I was brought up in a city by a mother who was very culturally sophisticated but didn't have an inherent connection with nature. She felt uneasy in long grass or in woods and was intimidated by bugs and insects. We as a family generally kept our distance from wildlife: flies were a nuisance, spiders were creepy and bees were to be avoided. Schools in those days did little to bridge the gap (thank goodness things have changed), and even though I loved being outdoors and read the *National Geographic*, loved our pets and was active outside, I grew up out of touch with the natural world.

My time with Christo brought me closer to the wild – he loved beetles and was very knowledgeable about birds and butterflies. Another influence came from my wife, Amanda Ferguson, who I met at Wye all those years ago. She was reading Animal Sciences, studying mostly farm animals but also with a particular interest in birds and bugs. Sometimes love brings with it a great focus – and direction – and I was intrigued by her interest and before long drawn into her world. Amanda gave me a different perspective on habitats. Much of conservation was scorned in those days, especially among the agriculture students, who tended to look on it as green nonsense. But when you think that 98 per cent of our species-rich ancient meadows have been lost in this country since the Second World War and consider the associated damage done to wildlife, you soon realize the importance of habitat protection. Life in a meadow is rich: by this I don't just mean full of impressive beasts such as grass snakes and wasp spiders, but with a cocktail of bugs and beetles, bees and flies, chirping crickets and grasshoppers galore, shrews and voles, slow worms and butterflies and moths – all of it part of an intricate web of life. If you compare this with a patch of intensively managed lawn, which has had every bit of life squeezed sprayed and manicured out of it, you begin to have a greater appreciation for long-grass habitats.

The wasp spider (*Argiope bruennichi*) makes a large round web typically decorated with a prominent zigzag pattern. The web floats among the flowering stems of meadow grasses, ready and waiting for any unsuspecting insect.

SPREADING THE WORD

I kick myself for starting late but thank goodness that initial obsession with plants took me down an avenue which opened up a world. A few key people influenced me in important stages of my life. Now, thirty years on and in charge of Great Dixter, I feel a responsibility to take full advantage of the way we can influence the lives of others. For the last four years Great Dixter has employed an education officer who engages with schools and families, hooking children into nature. Pond dipping, meadow sweeping for insects, moth trapping, butterfly netting, bird watching, walks across the meadows, time in our ancient coppiced woodlands, building fires and making dens, as well as growing vegetables and picking flowers, keeps everyone close to the land and for some opens up new horizons

Places like Dixter influence adults as well. A well-known garden celebrated for its vibrant and colourful borders, we have the privilege of a large audience. Still you can see visitors bemused by the wilderness, especially the long grass, just as I was on my first visit. Some feel so strongly that this naturalism is out of place that they demand their money back, leaving their disapproving comments in the ticket booth. This negative emotion is heightened especially during the months of July and August, when we have fewer flowers and more seed heads. Educating grown-ups is not as easy as tackling children, since older views are more set, but efforts through leaflets, tours and lectures as well as showing them a beautiful meadow

in its season makes a difference – things have got better certainly in the last few years.

The common blue butterfly (*Polyommatus icarus*) is widespread on grassland, although it populations have declined in recent years. Plenty are to be found on the Great Dixter meadows, feeding on bird's-foot trefoil (*Lotus corniculatus*) and black medick (*Medicago lupulina*).

Of course the majority of the visitors love it. They appreciate the vegetation moving even in the lightest breeze, the ever-changing tapestry, the summer sound of chirping grasshoppers and crickets, the drowsy low hum of bees, the fluttering of butterflies, the silhouettes of seed heads cast on to the York stone when the sun is low, the smell so evocative of the past, and the special atmosphere it creates of being at one with nature, and being part of a bygone age of a patchwork landscape of hay meadows and arable fields. All of this is heaven. Only yesterday two visitors, both motorcyclists clad in leather, their helmets held by their sides, were standing at the topiary lawn – a formal lawn fifteen years ago, now blissfully teeming with orchids, buttercups and ox-eye daisies. Looking across the dark green topiary pieces in yew floating above the long grass, one exclaimed with excitement to the other – 'Now that is what I call a lawn – we could do that at home!' A glorious meadow, a magical moment and a stirring of emotions: being there at the right moment can trigger something in all of us.

STARTING AT DIXTER

I kept in regular touch with Christo for a number of years after Wye before he offered me the job of head gardener. Dixter has always been magical but it needed an injection of energy. The place was tired and Christo in his seventies was already doing too much. We worked well together – he was my mentor and I his student. Much of my time was spent on the flower borders but we also wanted to turn our attention to the meadows. Christo made a push to tweak certain areas of long grass. Some were too coarse and lacked the diversity he wished for, dominated by vigorous grasses where the ground was too rich or smothered by reeds in the wettest places. In other instances he wanted to add to what was already there, visually strengthening the communities, adding more colour or another season or extending the groups. Christo wasn't averse to introducing exotic non-natives into his grassland and together we experimented. Much of this work is ongoing; some of our efforts have rewarded us immensely, but in other cases our trials have failed. In essence the meadows at Dixter continue to develop at their own pace and in their own way.

THE HORSE POND

I started as head gardener at Great Dixter in 1993 and one of the first meadow projects was to tackle the rush problem by the Horse Pond, a semi-natural pond to the east of the garden. With Perry Rodrigues (who is now Business Manager) we laid metres of perforated pipe in trenches of pebble beach in a herringbone pattern across the whole area, draining the land into the Horse Pond. This was done by hand, carting tonnes of grey sodden clay away on wheelbarrows, then loading it on to a tractor trailer. The aim was to make the area too dry for the reeds to dominate. And for a number of years it made a significant difference: the rushes dried up and the grasses moved in. But now the pipes have silted up with our very fine clay and the vegetation has returned to its natural state, dominated by rushes.

Thinking back now it seems naive of us to force a change on this wet area when we had so much other meadow elsewhere. This area in particular offered us diversity and a different habitat. Meadows are never uniform anyway; the composition conspicuously varies with the conditions, with individual species having different preferences and tolerances of climate, soil, drainage and nutrients. In terms of diversity and interest, it is surely better to embrace the variations within a site, each with its own distinctive community.

Had we been successful in draining this area we would have lost the rushes but also our colonies of water dropwort (*Oenanthe crocata*) and lady's smock (*Cardamine pratensis*). The dropwort is a beautiful umbel with its domed heads of off-white flowers, elegantly spaced yet bold, on tall, well-furnished plants. Reputedly it is the most poisonous of all the British natives, its potency concentrated in its white carrot-like root, with fatalities recorded in cattle when the roots are exposed during flooding or ditch clearing. I'm rather glad of its presence at Dixter, colonizing in waves along these wet patches and playing host to a stream of insects. I'm also glad to have an ever-expanding colony of lady's smock here, as it is the food plant of the orange tip butterfly (*Anthocharis cardamines*). The Horse Pond is only 360m/400 yards away from the Upper Moat. Both are on clay, both are wet, but they have differing vegetation, with the latter being dominated by meadowsweet (*Filipendula ulmaria*); I find this sort of difference fascinating.

Flanked by a dry bank shaded with trees on one side and a wet ditch on the other, the Horse Pond has its own special atmosphere. The damp ground is often waterlogged during winter months, supporting a meadow community different from anywhere else.

I know why Christo was nervous about the patch of rushes, because he felt they could encroach into adjacent areas, but twenty years on they are still contained within this bowl of wet ground, their perimeter and density fluctuating with the amount of rain from one year to another. This area has its own character, soft underfoot, with a slightly anaerobic smell, and is cool under the shade of the trees. Just a few feet away, where the ground rises and is more open, the vegetation changes dramatically, giving us precious diversity within a relatively small area. This reaction to conditions, ebbs and flows, is what makes meadows fascinating. On a recent visit to Newborough Warren on the southernmost point of Anglesey, for example, we moved from grazed fields to open fields of meadow, to areas that were damp and supported wild angelicas, meadowsweet, and water mint, to grassland that merged into woodland, to dunes of marram grasses (*Ammophilia arenaria*), sea holly (*Eryngium maritimum*) and a sea spurge (*Euphorbia paralias*). There was such a variation within a relatively short distance, supporting a rich biological community, and the whole place was alive with insects and birds. Dixter too has its variation and with it comes a richer community. So: I am glad that our pipes silted up and the vegetation took its own natural course.

BULBS IN THE MEADOWS

In other areas of the Dixter meadows, using Nathaniel Lloyd's bulb planter, we added thousands of *Crocus tommasinianus*, *C. biflorus*, and *C. chrysanthus* hybrids, planting them in large irregular sweeps along the front path leading to the house and in the main orchard just below the Long Border. Most of these have done well and have increased, except where the land gets flooded from time to time at the front of the house. The patches near the Long Border make a delightful display in February and March and have gone on to hybridize and seed themselves about. Christo loved all the different variants that occurred as a result of this crossing. We've stopped adding to this patch now, hoping that the colonies will go on to do it themselves. The later-flowering large Dutch hybrids (*C. vernus*), so adored by early bumblebees, are also strong here. They have continued to plump up and spread themselves by seeding down the hill, making irregular patterns of white, purple and mauve.

Also happy are the colonies of European dog's-tooth violet (*Erythronium dens-canis*), that we've been religiously dividing and increasing along the shady side of the Horse Pond. For a good many years clumps have been lifted, broken up and spread throughout the long shady bank, below which sits the area of damp ground adjacent to the Horse Pond. Again, most of the work was done using our old bulb planter, taking

The Dutch *Crocus vernus* hybrids have happily self-seeded in the long grass at Great Dixter for many years, making large colonies of predominantly purple, mauve and white.

We have encouraged the European dog's-tooth violet (*Erythronium dens-canis*) to spread throughout the semi-shade of the Horse Pond meadow. In the spring its chocolate-mottled leaves and dangling flowers rapidly appear out of nowhere catching you totally unawares.

out very neat plugs and placing the small, white, pointed bulbs in the holes before replacing the turf plugs. To start with our efforts hardly made an impression, but as time has gone by these bulbs have fattened up and the effect is stunning. Since Christo's death the area covered by the erythroniums has more than trebled, stretching all the way along and up the bank separating the Horse Pond from the prairie, making quite an impact as you drive into Dixter in early spring.

Along the same patch of ground, primroses and little *Narcissus minor* have also multiplied, making a lovely mosaic of yellow and lilac with the erythroniums. The softer lemon-yellow hoop petticoat narcissus (*N. bulbocodium* subsp. *bulbocodium* var. *conspicuus*) has more than quadrupled its colony, with the wiry dark green leaves of self-sown seedlings now seen stretching several feet away from the original place Christo planted them at the base of a large oak tree.

Christo was very fond of these miniature daffodils. He coveted the colonies of *Narcissus cyclamineus* at the RHS gardens at Wisley and was given a few bulbs by David Jewell. Only 25cm/10in high, with bright, pure yellow flowers with distinct windswept petals and a long, thin, protruding trumpet, these were planted in semi-shade on a little hump surrounded by wetter ground and they have increased rapidly, seeding and spreading throughout the area. In fact they are so successful

The pale form of the hoop petticoat daffodil (*Narcissus bulbocodium* subsp. *bulbocodium* var. *conspicuus*) has seeded prolifically around the large oak tree that flanks the Horse Pond, preferring to colonize drier mounds among the oak roots rather than venturing out into wetter areas.

that I fear for the safety of the hoop petticoats on the adjacent hummock, which are in danger of being overwhelmed. *N. cyclamineus* is not only more prolific and physically robust, it also dominates the space visually with its vibrant strong yellow flowers. So this year I may move the cyclamineus into another part of the garden where it can romp away unhindered.

EXPERIMENTS

Meadows are managed areas of wild and Christo loved experimenting with non-natives in grass. Just up from the Horse Pond there was dumping ground for his garden waste from the High Garden. Years of debris resulted in a bit of coarse ground, where the grass grew rank owing to the rich soil and the shade cast by overhanging Scots pine trees. Christo and I tried to establish oriental poppies (*Papaver orientalis*) here but never with success; they survived for a year or two before fizzling out. As the trees have closed in more this area sits in deep shade now so I'll have to think again – maybe a different erythronium would stand out in the spring?

The giant fennel (*Ferula communis*) planted in the long grass of the Cat Garden has also been a disappointment. After twenty years it has barely got above

15cm/6in high – the competition from the grass simply being too much. This doesn't mean that we give up; the fennels can be tried on ground where the grass isn't so dense – perhaps somewhere sunny and stony such as where the grass meets the edge of the coach park. I'll also find a home for the oriental poppies – in a less shady spot, still in fertile ground where the grass grows tall but is more open at the base.

A recent experiment with a new *Angelica* species has worked very well, with young plants already thriving in the boggy hollow of the Cat Garden, flowering with large plates of lime green in June. The white form of *Galega orientalis* has been another success, romping away in a troublesome patch dominated by ryegrasses on the edge of the Prairie. Pale sky-blue chicory has also done well here and the two look handsome together. Both are thugs and love the lush ground rising high above the tall sward that surrounds them. Hay rattle (or yellow rattle, *Rhinanthus minor*) has never taken here – surprisingly unhappy among tall grasses – and so the only option was to find plants that would cope with the competition. The galega could be troublesome and get out of hand but for the moment it's behaving itself.

THE PRAIRIE

A few years ago an inspection pit was dug right in the middle of what we call the Prairie, not too far from the galega patch. This area was an old fruit orchard but now only has trees around its perimeter. The grass here is predominantly bents and fescues, with a nice mix of knapweed, bird's-foot trefoil, and orchids. The excavations left an awful mess, so just out of curiosity I heeled in a creeping herbaceous spurge, *Euphorbia* × *pseudovirgata*, to see if it would cope with the grass. Having proven itself as a thug in the garden, making its way through the borders with slender stems, grey-green leaves and vivid acid lime-green flowers, it was likely to succeed here too, but then you never know. Taking a few years to settle, it gradually crept outwards and now occupies an area over 20m/65ft across, flowering in May and carrying on until September, colouring up into a body of orange and salmon in the autumn. This spurge was a solid vibrant mass for a few years but with age has opened up, allowing the other colours to drift into it. The lime-green mass looks particularly effective with the purple tufted vetch (*Vicia cracca*) and lesser knapweed (*Centaurea nigra*) both flowering among it in July and August, all with a backdrop of biscuit-brown grasses.

This area of the meadow where the spurge is was Christo's playground in which he experimented with North American prairie plants. Over a number of years he planted several out, but most of them hated our xdense British turf, which is alive, green and active over the winter months, unlike the brown winter-dormant sward that exists in the North American prairies. Ironweed (*Vernonia crinita*) and milkweed (*Asclepias*

RIGHT, ABOVE Named the Prairie, the patch of grass above the High Garden was home to many of Christopher Lloyd's meadow experiments, predominantly involving North American prairie plants.
RIGHT, BELOW *Euphorbia* × *pseudovirgata*, a long-season spurge with an adventurous rootstock, has made itself comfortable here in the Prairie, now forming quite a colony, all grown from a single plant heeled in here several years ago.

tuberosa) gave up the ghost pretty much from the start, *Silphium perfoliatum* gradually declined although a few colonies hang on in the semi-shaded areas. *Silphium terebinthinaceum*, the prairie dock has stood still, a couple of paddle-shaped leaves poking boldly out of the sward but only flowering twice in the last ten years. *Veronicastrum virginicum* 'Album', on the other hand, has continued to increase, making impressive clumps flowering in July. *Helianthus grosseserratus* has been another success story, as has *Eryngium yuccifolium* – ever dramatic in its ghostly presence. The pampas grasses have fizzled out, and *Eupatorium maculatum* Atropurpureum Group only hangs on along the semi-shaded hedge line where the grasses are more open. Where the grass is thick and tight these larger perennials struggle, because they are not able to cope with the early activity of the European grasses.

This area is usually cut in September, after the late perennials have flowered. This late cut encourages knapweed, which seems to be more prolific at Dixter since we have delayed cutting more and more, as we wait for the insect activity to slow down: this in turn allows the knapweed to set its seed and spread. Lesser knapweed dominates our meadows in August, when it is alive with bees and butterflies so we are reluctant to cut it down prematurely. Dyer's greenweed (*Genista tinctoria*) is also abundant here – grown from seed introduced from a local roadside verge. Now there are fat clumps throughout the area spreading by seed but also vegetatively. Also prolific are the orchids – twayblade orchids (*Listeria ovata*), which I believe are rhizomatous, are making large colonies especially in a triangular strip by the fruit cage. Common spotted orchids (*Dactylorhiza fuchsii*) have also made particularly dense colonies here, thriving throughout the area regardless of sun or shade. As the hay rattle moves through the sward, opening it up, the wildflowers follow, changing the colour patterns throughout the sward from one year to the next.

THE TOPIARY LAWN

One area of meadow that has developed dramatically since Christo's death is the Topiary Lawn. I remember the moment he decided to let the grass grow in what was once a formal lawn. I was on nursery duty, it was Sunday morning, and Christopher was sitting with a friend, Graham Harvey, on the Lutyens-designed family pew (which seats at least twelve), which looks out on to this area. The lawn had not been looked after in the proper sense, having been mown regularly but not given much in the way of feeding or weeding. It was under stress, covered with moss in the shady corner down by the nursery and riddled with broadleaved weeds (this of course came to work in our favour). Christo stopped as I was walking up from the nursery to check the pot displays and told me what he wanted to do (it was Graham's idea). At that moment I couldn't quite see that this was an act of brilliance, but I knew well enough not to doubt him. He could easily see that this would be a quick turnaround. We got on our hands and knees and on inspection we could see dandelions, clover, yarrow, bird's-foot trefoil, self-heal, ox-eye daisies and buttercups mixed in with the fine lawn grasses. It was all there but wasn't allowed to develop because of our constant mowing.

The family pew overlooks the Topiary Lawn, which Christopher Lloyd transformed from a formal lawn into an increasingly rich tapestry of wildflowers, including ox-eye daisies and many orchids, with the majority of the newcomers having self-seeded.

We decided to let the grass grow without any cultivation or pre-preparation, to see how things progressed. The first few years were dominated by white clover, buttercups and lawn daisies. Then gradually red clover took over from the white, its rich carmine-red flowers looking magnificent among the grasses and buttercups. Three years in and to our excitement we saw a common spotted orchid. Within four years we had five, and then by year six I counted fifty-six in a straight line from one end to another. Now, almost twenty years on, there are thousands.

Of course, I fully understand that success isn't solely gauged by orchids. It's exciting that we have such dense colonies but, more importantly, the long grass here is a complete antidote to the formality of the space, with a striking contrast between the free and mobile meadow and the dark-green static yew pieces rising dramatically from it.

OVERLEAF The dark green yew pieces seem to float in a sea of vegetation here – the perfect juxtaposition between the wild and the formal. The Topiary Lawn has improved dramatically with age, but it still remains contentious with some garden visitors.

The floral tapestry here has markedly improved with age, becoming more complex, with ox-eye daisies, vetches and meadow cranesbills. We tried adding other ingredients – plugs of *Dianthus carthusianorum* and harebells (*Campanula rotundifolia*). Both thrive in dry nutrient-poor grassland rather than our soggy clay, and our attempts met with failure; but it was worth a try.

Other than these two, the Topiary Lawn meadow has been a resounding success, with the yellow or hay rattle opening up the sward for orchids, bird's-foot trefoil, and ox-eye daisies and enabling them to make dense colonies. The grasses themselves, especially when brown, visually sit well with the dark green yew. The common bent (*Agrostis capillaris*, formerly *A. tenuis*) is abundant here, casting a light pinky chocolate haze over the ground and creating magic moments when the dew hangs off the grasses on early summer mornings. As the summer progresses shades of russet and brown dominate, with the grasses setting seed before the meadow is closely cropped for the winter.

Mixing topiary and long grass does not appeal to everyone. Some feel that topiary should only sit in a formal setting, as this is what it was designed for. And I can't say our topiary pieces are altogether happy, because their feeding regimes have been reduced so as to favour the nutrient-poor meadow. As well as this, the bottom of the topiary pieces, especially on the shady side, can become bare as the tall grasses cut out the light. But, personally, I think Christo created a patch of genius: a perfect balance between formal and informal – the sort of thing that takes your breath away when you see it for the first time. Coming in through the gap in the hedge by the nursery you suddenly enter a wonderland with the house sitting majestically in the backdrop. After all, there are enough gardens where topiary comes out of nice neat lawns – why not accept one where it comes up through a meadow?

The Topiary Lawn is a perfect room on its own and unlike any other space in the garden. Shaggy and wild when the grass is long and the topiary uncut, it dramatically takes on a more formal appearance in the autumn and winter once the meadow is shorn and the yews are clipped.

Hemi-parasitic yellow hay rattle (*Rhinanthus minor*) has been a useful tool in opening up many of our meadow areas dominated by grasses; once these are weakened other plants have the chance to seed in and establish themselves.

TAMING THE GRASSES

The introduction of the hemi-parasitic hay rattle (*Rhinanthus minor*) has had a dramatic impact through all the meadows at Dixter. So much more of it exists now than in Christo's time. It adds another texture and colour with its cool softer yellow, and is also a handy tool to keep the grasses at bay. Living on a boom-and-bust cycle, it is so effective in weakening the grasses that it eventually leaves itself little to feed on, causing its own population to crash. This allows other meadow ingredients to establish themselves before the grasses come back, with the rattle following in their footsteps. So, in effect, it creates a dynamic mosaic system which has ebbs and flows of grass density encouraging a varied community to build up. The downside to this is that the hay rattle dominates the finer grasses like the bents and the fescues and so dilutes the show from these. And I have also found it slow to get stuck into the tougher grasses, such as ryegrass, which is exactly where you want it to be effective. If you give it this tougher option the rattle gradually gets the job done but it can take ages before you see a significant difference, with results being erratic. On the rich ground below the Exotic Garden where *Inula magnifica* towers above everything, vigorous grasses dominate. It has been ten years since we first introduced rattle and only now are we seeing effective results, with the area more manageable, the grasses seldom lodging (falling over under their own weight) and with a greater diversity in the sward.

There are other areas that have benefited from the introduction of hay rattle. The lush ground by our burn pile at the bottom of the orchard is now certainly more diverse and colourful as a result, with the hemi-parasite weakening the grasses and allowing other ingredients to establish.

Since Christo's death I have also introduced more meadow cranesbills (*Geranium pratense*) into the lusher areas as they seem to cope rather well with the competition, giving much-needed colour late on. This has been a resounding success, especially along the edge of our water reservoir, a fertile bank dominated by *Phragmites australis* (the common reed), and in June the grey-green foliage of the reeds set off the blue flowers of the geraniums. The cranesbills also thrive in the semi-shaded areas in and around trees in the orchard.

NEW WORK AT DIXTER

Christo is no longer with us but our work with meadows continues. There is already a remarkable and complex mosaic system here at Dixter, where there are over a dozen different meadow habitats, and with it much diversity. The trick is now to manage all of this in an effective way to maximize their flowering and biodiversity. New meadows are being created on the estate all the time. A large field just below the garden in the non-public area was taken in four years ago. This used to be leased to Mr Young, the farmer next door, who grazed Sussex cattle on it for years. When Mr Young sadly died we took control of the field and let it turn into long grass. Our initial attempts gave poor results even though we strewed hay from the species-rich meadow sites within the garden. Hay strewing involves cutting hay from a seed-rich donor site and spreading it on the recipient site, covering approximately twice the area. This can give good results but there are so many variables. Cut the donor site too late and most of the seed drops while being cut and you also favour late species like knapweed. Spread the strewings too thick and the grass ferments and heats up as it rots, giving poor results. Shake the hay up too much during transport and you lose too much seed. Fail to prepare the recipient site well enough and the existing vegetation swamps out the germinating seedlings. Sometimes traumatizing the ground to make way for the new seedlings allows invasive weeds already present in the seedbank to establish.

In this case the problem was the heavy thatch – that is, the tussocky dead material built up at soil level which cuts out the light and does not allow germinating seedlings space to develop. So, we chain-harrowed the whole area, scratching away this inert layer that had built up among the grasses over a number of years. The whole field turned brown as Bob Common, our eighty-year-old tractor driver, zig-zagged the harrow across, ripping the ground and tearing out the thatch. The dead material was then raked up and taken away, leaving a stressed field with plenty of bare ground ready to be strewn with hay from various orchid-rich sites. The job became particularly enjoyable as members of the Northiam Horticultural Society got involved. It was thrilling to see slow-worms, to handle hay, to engage in communal work on the land, spreading the load around as it was thrown off the trailer, knowing that we were putting life back into the area.

This little project was a pleasurable joint effort that brought the community together, sweetening further the very positive results. Within a year this field, previously

A young Aysegul Garrett is helping a group from The Northiam Horticultural Society strew seed-rich hay from the meadows at Great Dixter on to the species-poor pastureland adjacent to the garden and Dixter's ancient woodlands.

made up of 95 per cent grasses, became 50 per cent grass and 50 per cent wildflowers. Today, three years later, there are orchids, ox-eye daisies, hay rattle, buttercups, knapweed and red clover. Only about 30 per cent of the field is grass now; the rest is rich in flower and rich in life. Hay rattle has taken a hold and moves through the area opening up patches. Ribwort plantain (*Plantago lanceolata*) seems to follow the rattle, with ox-eye daisies next to spring up, and lesser knapweed thrives here. Common spotted orchids have now started to appear already, with over twenty counted this year. I want these to continue to spread, so our cutting will be delayed until they have ripened their seed (usually in the second week of August). The big question now is: do we create something different from all the other meadows at Dixter by adjusting the management of this field? Perhaps as woodland borders this area on two sides we should blur the lines between the two. Woodland and meadow would have co-existed in ancient times before we enforced the separation. Allowing the field edges to grow, encouraging thickets of scrub, brambles and blackthorn to flourish, varying the grassland habitat from open to semi-open and then to shade will give us a mosaic of scrub quite unlike anything else that exists on the estate. It would be interesting to turn the field into pasture for the duration of the winter by using cattle – allowing them to lightly poach the land, their hooves opening pockets which would encourage germinating seedlings.

Ribwort plantain (*Plantago lanceolata*) is often first to colonize ground left bare as a result of hay rattle weakening the grasses. As other plants then seed into this space and the grasses come back, the plantain becomes less dominant.

LESSONS FROM EUROPE

We could adopt the complex system of agriculture seen around Transylvanian villages, for instance. There, some areas are kept for hay, others grazed regularly, some grazed only in winter, some cultivated, some rested; using a range of livestock and a range of methods produces a complex matrix of sward structures, resulting in great diversity. To a lesser extent we can mimic this process on our estate at Dixter, monitoring areas before the transformation and then throughout the process to give valuable comparative data. This work as we go into the future is unbelievably exciting.

Perhaps cattle could be used to graze some of our fields at very low density throughout the year, allowing them access into a part of the adjoining woodland. This would follow the naturalistic grazing pasture processes favoured by some

Throughout Europe terrain that is too steep or too rocky has escaped the onslaught of intensive agriculture. These areas are often grazed at low density or hayed and remain the perfect haven for wildflowers and meadow communities.

conservationists in Europe. The pioneering work on this is the Oostvardersplassen project in the Netherlands, where over two decades ecologists have been researching the influence of large herbivores (directly or indirectly) on the ecosystem. This is a dynamic approach where the natural processes are driven by the grazing of semi-wild herbivores with little management. In Oostvardersplassen over 2000 hectares/5000 acres of grassland has been reclaimed and developed this way, resulting in a shifting mosaic of vegetation dependent on stocking densities. For many of us, hay meadows and intensively grazed pastures are at the opposite ends of the spectrum and this approach would be where the two meet, along a wood-pasture system. And goodness knows what this new management will bring, but we do know that varying of habitats and their management is bound to favour different swathes of floral and faunal communities and this has to be a good thing as far as biodiversity is concerned.

CUTTING AND GRAZING

The estate at Great Dixter already has a complex management plan and adding to this will be easy. Currently, some areas of our mixed woodlands are cut on an annual basis, scalloping the edges of the rides to allow light and heat in for insects, whereas some of

the woodland is cut on a twenty-year rotation and some of it is allowed to develop into high forest. In the same way some of the grass is cut once a year, some of it twice, some areas are cut and grazed by sheep, some cut once every two or three years. All of this gives us variety.

Were it not for the close proximity of our ornamental borders we could use sheep to graze down our meadow areas once the long grass has been cut and collected. However, electric fences have proven ineffective in containing sheep that would much rather eat the lush vegetation of the garden.

In terms of management, even the way we graze and what we graze with will make a huge difference. Grazing some areas rather than cutting is important because it creates patches of tall and short vegetation, adding structural diversity – which is of great importance in encouraging insect diversity. The habitat requirements of insects vary: for instance tall, stable swards favour web-spinning spiders, while shorter swards are ideal for grasshoppers and crickets. Heavily poached areas allow certain insects to bask in the sun and dung is important for supporting its own community. Also, different species of livestock will have varying browsing and dunging behaviours. This is bound to influence the species diversity in a sward as well as the structural diversity. Sheep bite the vegetation, grazing it right down producing very short swards. They avoid tall growth and tussocky areas, and often leave dead material including grass stems. They can select material very low in the grassland profile and can pose a threat to rare plants if these are sensitive to grazing. Their droppings are small and have little effect on the sward structure. Cattle and horses make large soiled patches and avoid further grazing around them, which creates a lot of structural diversity. Horses graze in a similar manner to sheep, biting the vegetation close to the ground and producing short swards, even shorter than those produced by sheep. They are, however, heavier and will poach the land more. They will take tall grasses and tussocks and, like sheep, can be selective, squeezing out vegetation that is sensitive to grazing. Cattle have different dentition and cannot graze as closely: they bite, pull and tear the vegetation, thereby maintaining longer swards. But they will graze tall grasses and tussocky areas, including grass stems, and will also take dead material, including leaf litter. And, of course, stocking density will also have an effect.

CONSERVATION

For many ecologists the current state of affairs with meadows is one of doom and gloom. You still see a few fields in Britain maintained in the old-fashioned way but most have been lost. Hedgerows have been ripped out, wetlands drained dry, vegetation sprayed lifeless; fertilizers have not only polluted our waterways but have had an impact on diversity within our grasslands. Meadows have been ploughed up and over-seeded with vigorous grasses, and one piece of land has been detached from another, losing physical connections that our wildlife needs. Studies have shown that adding only 50 kilograms of nitrogen per hectare has a profound effect on the number of herb species present, in certain cases reducing the number by half. This is certainly due to gross feeders out-competing the slower-growing species. Also the practicalities of making silage or haylage – the time and skills needed – have made haymaking a redundant farming practice, resulting in many hayfields being turned into short homogenous grassland, which is a poor habitat supporting very little wildlife. The popularity of summer cutting does not encourage prolific insect life and in fertilized fields the resultant long vegetation shades out insects that require lower, more open swards.

Research carried by Plantlife (a conservation body set up in 1989 concerned with the protection and conservation of wild plants) reveals that since records began in the seventeenth century eighty species of flowering plants, mosses, liverworts and lichens have become extinct in Britain. But the situation is worse than this, as we cannot judge the state of affairs on extinction alone. The relative abundance has also declined, as many species commonplace in former times now occur in low numbers in isolated patches and in fragile communities. More than nine hundred species of plants and animals have been given conservation status because of the severity of the threats facing them. Habitat loss has been a key factor in all of this. The intensification in the management of lowland wet grassland, for example, has been cited as one of the key reasons for the decline of breeding waders throughout Europe. Cutting early has become increasingly popular and this has meant the loss of a valuable breeding habitat. Drained, monostructured and improved swards have created a massive reduction in the number of invertebrates and this has a knock-on effect on wader chicks, which require invertebrate-rich habitats to thrive.

OVERLEAF A few years on, and the field hay strewn by Aysegul and company has exploded into a community of buttercups, hay rattle, ribwort plantain and an ever-increasing colony of common spotted orchids (*Dactylorhiza fuchsii*).

Many consider the attempts to conserve species-rich grassland to be failing – despite tremendous efforts – owing to continued land fragmentation and agricultural intensification. Ploughing, draining, weed-killing, fertilizing and seeding have led to a monoculture of intensively farmed grasses. But the unimproved meadows that remain are a safe haven to wildlife as well as important sources of seed and a valuable resource for research.

LOCAL HEROES

Will the increased awareness from the current generation turn back the tide? This is difficult to predict, but we are already seeing more cases of public backlash when areas come under threat. Recently a local wildlife site in Bicester, Oxfordshire, was given reprieve from development after a successful High Court challenge. The developers lost their case because they failed to abide by the Environmental Impact Assessment Regulations; the verdict saved valuable areas of unimproved flood meadow supporting populations of the black hairstreak and brown hairstreak butterflies.

There is a strong meadow community fighting the cause. This group is made up of meadow-makers, gardeners, farmers, and conservationists. Keith Datchler, the manager for Beech Estate near Battle, East Sussex, has led by example for many years. A farmer-turned-conservationist, Keith is responsible for creating over 900 hectares/2000 acres of meadow in the High Weald. Most of his work has been at Beech Farm Estate but he has influenced many throughout our area. A farmer with a deep love and understanding of the countryside, Keith is someone who has taken the utmost sensitive care for the land under his watch. He embodies the saying: 'live life as if you are going to die tomorrow and farm your land as if you are going to live for ever.'

Recently I spent an afternoon sitting with him and his associate Dawn Brickwood of the Weald Meadows Project. We all realize that gone are the days of traditional farming but it is unfortunate that so much damage has been done to our land in the name of productivity. Keith refers to *Silent Spring* by Rachel Carson, written in 1962 and well ahead of its time, a hard-hitting book denouncing the adverse effect of humans on the countryside with particular attention to the detrimental effect of pesticides on natural ecosystems. Carson's message had a powerful impact on the environmental movement and left a significant mark on Keith.

The Beech Estate where Keith has done most of his work was bought in the 1950s by the industrialist Sir John Wills. The estate was intensively managed as a dairy farm, but in the 1970s Keith found himself in charge of a pair of old-fashioned homesteads, Rocks Farm and Bunces Farm, which had been bought by Sir John in addition to his existing estate. Isolated, and difficult to work, these had been in the hands of old-fashioned farmers with little input, development or improvement, ticking along on poor soils, with a considerable amount of the land locked up in hay production. The fields were never fertilized or sprayed and remained a landscape of a bygone era. Keith's initial reaction was to turn these farms round to something more productive. But for the first year he decided to 'shut up' the land (remove the grazing cattle to allow the grass to grow) before he finalized his plans. That June, Keith dropped in to check his hay crop and saw a sight that set his spine tingling and made the hairs on his arms stand on end. It was a spectacle that was a life-changing moment transporting him back to his childhood – fields rich with wildflowers, with orchids, buttercups, clover, ox-eye daisies and sweet vernal grass.

Keith grew up in London, with his family taking occasional trips down to his aunt's house in Deal. Here Keith would find himself immersed in the countryside, taking long walks in the surrounding fields. And even though he went on to agricultural college and was taught that productivity was king, the connection he made as a child never left him. In an instant his childhood memories came flooding back. With the backing of Sir John Wills and the onset of stewardships schemes encouraging landowners, the Beech Estate gradually changed its management to be more sensitive to the land and to be environmentally responsible. Instead of maximum productivity, the brave new emphasis was based on conservation.

Over the years Keith has increased the diversity on the farm by seeding areas from rich donor sites on the estate. Trial and error has refined his methods of improving the range of species in meadows. His initial attempts involved cutting and baling seed-rich hay from various parts of the estate and spreading this over areas that needed enhancing. These attempts gave mixed results, as using robust farm machinery knocked off much of the seed. To overcome this Keith developed alternative, gentler, cutting and collecting techniques. Power-harrowing, which traumatized the land, opening it up and making it more receptive, coupled with allowing the emerging seedlings breathing space to develop unhindered, gave him far greater success. He experimented and analysed, tweaking and fine-tuning all the time. Sometimes he used sheep to trample in the seed whilst keeping the existing competitive sward down. With his ability to judge the conditions and the practical knowhow to back him up, Keith has turned meadow-making into a fine art. As you would expect, his meadows at Beech Estate are among the richest and most impressive I've seen. And with decades of experience under his belt Keith continues to inspire, consulting on and making meadows throughout the High Weald. For his achievements Keith deservedly received an OBE in 2008.

Dawn Brickwood has been working with Keith for many years. With local Sussex seed firm Agrifactors, they form a group called the Weald Meadows Partnership (WMP). Like Keith, Dawn is from a farming background. After working with a dairy herd for over ten years she decided to take a Rural Environmental degree at Wye College, followed by a period volunteering for the Farming and Wildlife Advisory Group, and then joined the Weald Meadows Initiative (WMI) in 1998. We met her first shortly after this when she set up a seed trial at Dixter looking at germination rates of local provenance meadow seed. Well known in our area, Dawn has been key in advising a broad base of clientele about best practice in meadow-making. Keith has been linked with WMP since its pilot years in 1996, and together they bring farming experience and practical know-how as well as years of experience refining practical methods to give greatest success.

Now they are both playing a part in the Coronation Meadows project, a very welcome initiative by HRH Prince Charles in trying to create meadows in each county – aiming at 100 new hectares/250 new acres of meadow. Meadows desperately need support and having Prince Charles's name attached to a project adds power and gravitas.

SUPPORT FOR CONSERVATION

New meadows *are* being created all the time, but at the same time old meadows are being lost. Unfortunately, even now, the economics for landowners and farmers do not add up, since the grants available to support a less-intensive approach of farming are not sufficient to bridge the gap in an industry where profit margins are already low. But there is an increasing interest in biodiversity and some wealthy landowners are really making a difference by restoring large areas of grassland to meadow. There is also a welcome and reassuring awareness among the young, with enormous efforts from schools and education centres to make a difference. Many land-based study courses in higher education institutes include habitats and habitat management with a more sympathetic attitude to conservation. Certainly, sustainable farming is no longer dismissed as hippy nonsense.

The media are helping the cause too, with recent television programmes on wildflowers and pollinators and increased press coverage on the worrying plight of our invertebrate pollinators. The dramatic loss of meadow habitats over the last seventy years has had a huge impact on insect populations in this country. One example is the extinction of bumblebee species, with seven out of the twenty-four species found in the UK classed as under threat. This has triggered conservation bodies to act and, in the case of the bumblebees, the Short-haired Bumblebee Project was set up in 2009 by Natural England, the Royal Society for the Protection of Birds (RSPB), Bumblebee Conservation Trust (BBCT) and the conservation and research group Hymettus, with the aim of reintroducing the short-haired bumblebee (*Bombus subterraneus*), formerly extinct in the UK, and establishing a self-sufficient, self-sustainable population. The project is run by Nikki Gammons and is centred on the Dungeness and Romney Marsh area, as this was the last recorded territory of the short-haired bumblebee and also provides a home for five of the seven other threatened species. This location is ideal, as it has large expanses of protected Site of Special Scientific Interest land along with national and local nature reserves. Nikki's work involves active co-operation with farmers, land-owners and conservation groups in order to recreate suitable flower-rich habitats, and she gives advice on seed sowing, management and maintenance to maximize the results of their joint effort.

The *Bombus subterraneus* queens are reintroduced to the UK from populations in Sweden and released on the RSPB Dungeness Nature Reserve where there are ample feed crops such as red clover (*Trifolium pratense*), ground ivy (*Glechoma hederacea*), foxgloves (*Digitalis purpurea*), and white dead nettle (*Lamium album*). A large body of volunteers helps with monitoring, regularly surveying across the release zone to assess the progress of the newly released bumblebees. Habitat improvement has been a key objective – the project currently has over 850 hectares/2100 acres managed for flowers, with Nikki aiming to recreate another 20–30 hecatres/50–75 acres of flower-rich ground. It's early days yet but the improved management of flower-rich zones across the release area has shown a positive effect on the abundance of the five rarest bumblebee species found in this area.

A host of bees clustering on echinops flowers: with the sheer variety of flowers in our gardens, meadows, and woodlands, along with a range of habitats and sympathetic management, Dixter is rich in biodiversity.

PLAYING OUR PART

With our high populations of red clover, bird's-foot trefoil (*Lotus corniculatus*), and lesser knapweed (*Centaurea nigra*) in our meadows, as well as the flower-rich garden, Great Dixter seems to provide an ideal habitat for bumblebees. Nikki Gammons regularly surveys the site, along with Sarah Seymour and Catherine Haydock, who have both been involved in the biodiversity work at Great Dixter. Their monitoring has shown that eleven species exist on our estate, including some from the endangered list such as *Bombus ruderarius* (the red-shanked carder bee) and *Bombus ruderatus* (the ruderal or large garden bumblebee). The presence of these indicates that the habitat also supports many other pollinators: honey bees, hoverflies, flies, beetles, moths, and so on. We are starting to monitor these at Dixter with the help of specialists who give their precious time and expertise to help with our surveys. More can be done to link with universities and research bodies for further investigation. The results show a very healthy habitat and will play a part in steering the right management path to take us forward.

There are many triumphs at Dixter – more than 380 species of macro-moth have been recorded to date on the estate, and more than 150 species of spider.

Some of Great Dixter's meadows are cut as late as mid-September, favouring later-flowering species such as lesser knapweed (*Centaurea nigra*), and encouraging the insects and animals that enjoy this useful late cover.

As well as this, our flower-rich meadows have great potential for solitary bees as well as bumblebees. Last year Ian Beavis, a specialist from Tunbridge Wells, found a banded mining bee (*Andrena gravida*) – a Red Data Book 1 (nationally endangered) species – on a patch of dandelion by Lutyens' circular steps. This bee historically had a stronghold in the High Weald but had not been seen for many years until Ian rediscovered it near Tunbridge Wells in 2011. Also found in our meadows was a female red-girdled mining bee (*Andrena labiata*), a nationally scarce species associated particularly with open grassland. Finding these on your doorstep certainly has a feel-good factor and we are driven to do so much more – from small-scale habitat piles to the larger-scale making of meadows, and to changing our management techniques in our woods as well as in our grassland to give us better results. We ought to survey more and feed the information into regional and national databases and certainly create large buffer zones where the grass and woods merge. Our whole estate, including even the car parks, can play an effective role in all this. It is also important to stand back and see how our land links with our neighbours – do we sit in isolation or are there ways we can link habitats together so that collectively they become more effective? Wildlife corridors needn't be made up of just meadows and fields: hedges, scrubland and woodland can all play a part.

Meadows strike a deep and nostalgic cord in most of us. Beautiful in their make-up, with an ever-changing picture of flower and insect life, they resonate with a feeling of aptness and harmony in the countryside.

MEADOW-MAKING

The fascinating factor in all of this is that each bit of meadow and each habitat is different – the dynamics varying from soil to soil and from site to site. Therefore it is not surprising that the methods people use to create meadows differ as well. I like speaking to landscapers and meadow creators: they have to get results, in all possible situations, often on a limited budget, perhaps with time pressures as well. Faced with adversity, they are forced to come up with the most practical – and often diverse – solutions.

For years I believed that the most successful meadows are made on poor soil, the low nutrient levels keeping the thugs at bay and allowing a more diverse community to develop. But if you have only one rich field then you have no choice but to use that soil. Of course you can take drastic action, scalping the topsoil and carting it away, leaving the more inert subsoil to put your meadow on. You could also deep plough, inverting the fertile topsoil under the less fertile sub-soil, or mask the rich soil by putting a hefty inert layer on top of it. These are all perfectly good practices and successful methods. But, in most cases you won't have the ability to change levels or bring in heavy machinery and will have to go with what you've got.

Colin Reader makes meadows around us in Sussex, and has developed his own techniques, proving to be very successful in creating meadows on large fertile agricultural fields. Colin is predominantly an ecologist working on wildlife habitat creation. Through a series of experiments (where successes were coupled with as many failures – and important lessons were learnt), Colin has come up with methods and seed mixes that give him quick results on fertile ground, overcoming weed problems and resulting in flower-rich meadows. The characteristics of Colin's wildflower landscapes are that they are very colourful, with a long flowering season, and include a range of wildflowers not normally available in commercially available mixes. His mixes are specially composed by him and contain twice the ratio of wildflowers to grasses, providing a colourful and species-rich blend. He also produces assortments for annual meadows, as well as permanent mixes for woodland, semi-shade, clay, loam, and sand, and compositions for chalk and limestone.

Colin is not averse to using chemicals, firmly stating that without them he could not have created some of his largest and most successful meadows. He strongly believes that you have to be pragmatic about what will create these favourable habitats, and that as long as chemicals are used in a responsible manner the good results will far outweigh any damage done. When faced with a weed problem Colin will spray the whole site with glyphosate. Once the vegetation is dead he cuts everything down. The field full of stubble is then left to green up as it grows and seeds germinate, then the field is sprayed again. If the site is very weedy this treatment may be repeated, and in the most troublesome areas the ground is left barren and the treatment carried out several times. The resulting dead debris is simply left behind, not ploughed in or scraped away, as this would result in disturbing the topsoil layer, exposing and activating the weed seed bank. The meadow seed is then sown into this debris layer, which not only provides an organic 'start-up' medium but also offers the seed physical anchorage, stopping it from washing or blowing away.

Creating a balanced meadow that has a multi-layered, dynamic community takes time, planning, skill, effort, and often some luck. For those who persevere and are successful the rewards are complex and profoundly gratifying to the senses.

On the most compacted sites where cultivation is needed but has to be avoided because of the weed seed bank, Colin covers the whole area with an inert layer of soil. He then sows seed directly into this, giving the germinating seedlings a sterile starting rooting zone before they hit the compacted layer. His seed mixes may have up to forty species, including some cornfield annuals mixed in with the perennials to give colour for the first year. Colin also has specific mixes for flowering lawns that contain clover, self-heal, wild thyme, lawn chamomile, and violets, all plants that will tolerate mowing on a regular basis. Areas sown with these mixes can be mown once every three weeks at around 5cm/2in, giving you a usable lawn as well as a flowering tapestry that not only looks pretty but has wildlife value too.

Since Christo's death there has also been much progress in the work of Dr James Hitchmough and Dr Nigel Dunnett from the University of Sheffield's Department of Landscape. They have continued to develop their methods and mixes, which are used to transform urban landscapes, creating nectar-rich plantings and giving derelict areas a real lift, bringing colour, scent and much-needed life into dull places including roundabouts and industrial sites. Their work has become very popular with local authorities and garden and landscape designers, and was a great success at the London Olympics where they created areas of impressionist, high-impact annual and perennial plantings which caught the public's attention and brought an appreciation of wildflowers to a wider audience. What happened in the Olympic Park and the way it was so positively received is a reflection of the growing trends.

A BRIGHTER FUTURE

Lately, I feel there has been a positive shift towards wilder planting. Meadows are very much in vogue in gardens and among garden designers; they also seem to be in favour with the general public. Just taking the village of Northiam as an example, there seem to be many more areas devoted to long grass. This may be to some extent the result of local council cutbacks, but I like to think that increased awareness plays a part too. I see more churchyards left long and quite a few public parks have sections devoted to meadow. Certainly there are fewer complaints about uncut grass in our garden, and many more questions about how to create meadows. But we all recognize that still more work needs to be done; there's plenty of opportunity to allow more verges to grow, especially in suburban areas. Sometimes when driving to work I see banks of white clover, buttercups, dandelions, and lawn daisies looking amazing – then along comes a contractor to mow it all down. It all seems so pointless and a waste of manpower and resources to create such a sterile environment. Although leaving the grass long has practical implications – it's troublesome to cut and has to be picked up – the right machinery is out there to make the job easier, and it could be budgeted for by forward-thinking councils.

Everyone needs to make the effort, as too much that is precious has been lost already. Of course the major difference to be made is in the working landscape where there are hundreds of thousands of acres. However, every little bit will help, however large or small – they will be like pieces in a jigsaw or corridors linking wildlife together and, in doing this, preserving the iconic British summer landscape.

It has been many years since this book was first published. Christo is no longer with us – he died in 2006 – and times have changed. Great Dixter is run by the Great Dixter Charitable Trust, governed by a positive group of trustees under the brilliant leadership of Geoffrey Dyer, our chairman. Biodiversity remains as important to the Trust as it was to Christo, and with this in mind we plan to take positive steps forward. The estate has much to offer and we have much to learn from it. Meadows sit at the centre of all this, as close to our hearts as they were to Christo's. Our work on this front goes from strength to strength, using all the bright minds out there to make a difference to our land. It is our hope that we will be able to inspire others to do the same.

A view of the orchard leading up to the famous Long Border and the back of the house at Great Dixter, clearly showing how the countryside is embraced as it flows into the core of the garden.

Introduction
to the first edition

To see a meadow in bloom is a great delight – it's alive and teeming with life, mysterious, dynamic and seemingly out of our control. The community of flowers and grasses is rich and colourful and so are the butterflies, moths, grasshoppers and other fauna that go with it. It is no wonder, with the biodiversity of our countryside fast disappearing, that meadow gardening has become fashionable. As a conservationist concept, meadows take us back to nature and away from the gaudy trappings of civilization.

Wild gardening in grass is set to become even more fashionable, now that the fine-sward lawn – that featureless expanse of green which is the icon of the English garden – is, officially, under threat from climate change (see *Gardening in the Global Greenhouse: Climate Change and Gardens* by Drs Richard Bisgrove and Paul Hadley, 2002). With drier summers we have already experienced longer periods of brown lawns and the cost of irrigation is escalating. Winter rainfall is on the increase, causing more compaction and moss growth, as well as problems maintaining fertility as nutrients are washed away. All of which is good news for many meadowland communities. Unlike turf, composed of bitty little grass blades, a meadow is constantly changing. There's shadow and movement; the grasses bend before the wind and the light catches and reflects their stems. Add other kinds of plants and you feel you are arriving at the essence of nature. And unlike a lawn whose appearance is so vulnerable to drought, bleached meadow grasses are magical, especially when laden with dew.

A meadow's serenity suggests lazy abandon but, like a garden, establishing and maintaining it is almost entirely dependent on human management. Achieving an image of nature as we might dream it to be is not as easy as it looks. While they save on the time and expense associated with traditional lawn-care regimens, rather than being a thoroughly low-maintenance option, as many gardeners have been led to believe, meadows need constructive hard thinking.

All over the world, natural grasslands were originally confined to areas where trees could not grow because it was too dry, wet or windy, or too high up. The so-called natural meadows that survive are generally a transitional stage between grassland mown for hay, or pasture grazed by farm animals, and the woodland that would naturally dominate. Given time and neglect, a meadow reverts to scrub and, ultimately, to woodland.

Meadows, then, are semi-natural vegetation. The plants are wild, yet the habitat has been created and maintained by human management (more of this in Chapter 2). Nowadays, however, a meadow has come to mean wildflowers growing in a grassy place that will survive being cut (or burnt) to the ground once a year. As we shall see, it is a way of gardening that takes many forms and is dependent on

A flower head of hoary plantain, *Plantago media*, with grass spikes.

all kinds of factors, such as climate, altitude, soil type, moisture, nutrient and light levels, and so on, as well as on the controlling hand of the gardener.

The most stable meadows, and those that have the richest tapestry of flowers displayed against the soothing coolness of a grass sward, are those made in old turf as they are in mine at Dixter. We have at least half a dozen areas of rough grass within the garden that range from flower-studded lawns and orchards to damp areas and an experimental prairie. The Dixter meadows are the subject of Chapter 2, as I thought it best to start at the centre of my little world, with what I know pretty intimately, and to work outwards from there. In Chapter 3 I consider the setting and siting of meadows before discussing, in Chapter 4, various methods of starting a meadow from scratch and aftercare of new meadows. Chapter 5 is about the management of established meadows.

Dixter's meadows harbour a beautiful assortment of grasses (Chapter 6), whose role in the meadow scene should never be underestimated. Chapter 7 and 8 are the longest in the book, since I need to describe the hundreds of colourful perennials and bulbs growing in our meadows, with many a sidelong glance at plants in habitats very different from ours at Dixter. There is no annual meadow area at Dixter, but I know of many gardeners who harbour a romantic notion of a field (or even a patch) of grasses ablaze with poppies and cornflowers, so how to achieve and maintain this ephemeral vision is the focus of Chapter 9, entitled 'The Cornfield Scene'. We have an experimental prairie at Dixter, which I use as a springboard to discussing North American prairies (Chapter 10) and the plants that thrive there. Chapter 11 explores wider issues – green areas in the urban landscape and their enrichment as meadows, and roadsides, which include a very large proportion of the relatively permanent meadow areas left to us.

LEFT A cornfield scene where herbicides are in abeyance and 'weeds' – field poppies, cornflowers and *Bupleurum rotundifolium* – can take over in the following year.
BELOW A cross-section of the lower orchard meadow at Great Dixter showing a bug's-eye view of the complex community of grasses, buttercups, hawkbit, clover and common spotted orchids (*Dactylorhiza fuchsii*).
OVERLEAF Ox-eye daisies and common spotted orchids mingle with bird's-foot trefoil (*Lotus corniculatus*) and meadow grasses. The meadow is home to a diverse microcosm of life which ebbs and flows with the seasons.

I. Nature, and the Hand of Man

While the inclusion of meadows in gardens is relatively recent, as an aspect of farming practice, they go back thousands of years, in Europe at least. In North America there is a meadow (as opposed to natural grassland and prairie) heritage, but it is even more recent. The richness of ancient meadows is due to the unaltered cutting, grazing or burning regimes imposed on them, continuity in management being important if species are to establish and thrive.

So what is a meadow and how does it differ from pasture and natural grassland? Strictly speaking, a meadow is an open tract of land where grass is grown to supply winter feed. The hay is harvested late in summer, allowing many of the plants to seed, and then the plot is grazed in autumn and winter. The feet of grazing animals disturb the soil, making ruts and bare patches that provide seeding opportunities without competition from mature neighbours. Pasture is grassland kept short by grazing animals for much of the year, thus giving flowering plants a less competitive environment in which to thrive. Natural grassland is controlled by the terrain (too high, too rocky and so on), acts of God – fire or flood, extreme drought, even strong salt winds – or wild animals, such as deer or rabbits.

The plants in pasture and meadow can be very different. Meadow-sweet, ox-eye daisies and salad burnet are examples of plants with long stems that cannot withstand summer grazing and are found in meadows. White clover cannot compete for too long in long grass and so is indicative of pasture, which has a greater incidence of plants with leaves in the form of rosettes, such as dandelions and daisies, as well as aggressive thistles and ragwort.

Nature, rather than man, played the largest part in the early stages, the natural climax vegetation in temperate latitudes being woodland. In Britain, for example, when the last Ice Age ended in around 12,000 BC, most of the country was glaciated, but south, roughly, of a line you could draw from the Thames basin to the Bristol Channel it remained unglaciated. Even so, trees could not get a foothold and a sub-arctic tundra grassland flora dominated. From 12,000 to 8000 BC, climate change enabled a predominantly woodland flora to become established, with pockets of grassland flora that were to be valuable reserves when changes in climate and circumstances allowed them to spread to other parts of the land. At this period the sea was much lower than now, so that Britain was still attached to the Continent, with the Thames most likely flowing into the Rhine. Between 7000 and 5000 BC, the climate warmed up to be even warmer than

This detail of The Unicorn in Captivity from *The Hunt of the Unicorn as Lover* (1495–1505) shows some of the flora and fauna in the lower part of the tapestry, with male orchid (*Orchis mascula*) revealed against the unicorn's white body. A butterfly alights on the carnation (*Dianthus caryophyllus*) and dragonflies hover over the blue iris and Madonna lily (*Lilium candidum*). A small frog lies among the violets above the A of the monogram in the tapestry's lower right-hand corner.

Lydden Down in Kent is managed under different grazing regimes to preserve its invertebrate and floral biodiversity. Tracts of chalk downland were among the earliest areas used for grazing because the thin soil enabled easy tree clearance. Typical flora may include salad burnet (*Sanguisorba*), chalk milkwort (*Polygala calcarea*), cowslip (*Primula veris*), dyer's green-weed (*Genista tinctoria*), horseshoe vetch (*Hippocrepis*), fragrant orchid (*Gymnadenia conopsea*), autumn lady's tresses (*Spiranthes spiralis*) and common centaury (*Centaurium erythraea*).

today. But we had a hunting, not a pastoral, population, so with little pressure on the land the vegetation cover remained largely unaffected by man. From 5000 to 3000 BC, the climate continued to be warm, though much wetter. The land was sinking, the North Sea formed and Britain was finally severed from the Continent. Once Neolithic man started practising agriculture, forest clearance began to make way for grazing flocks and herds. Chalk downland and limestone escarpments like the Mendips became the earliest grazing areas, because the thin soil enabled easy tree clearance. Regions with highly nutritious grazing, such as Romney Marsh, had never been covered by trees anyway, being insufficiently drained. As the numbers of grazing animals increased, so did the need for grassland. Agricultural pressures soon started changing the face of the countryside. Similar changes were happening throughout the northern temperate world.

During the Iron Age, 500 BC to AD 40, wood needed for fuel as well as building brought about even more forest clearance. Then, from AD 43 to 400 the Romans brought urban industrial civilization but, besides land for cattle grazing, yet more woodland was cleared to provide the grassland needed to support a flourishing wool trade that built up to the famous export wool trade of the Middle Ages.

After the Romans' departure nearly seven hundred years elapsed before the Norman conquest. Various invasions from the Continent occurred during this period, notably by the Saxons, whose settlements were always close to water supplies. They had waterside meadows for summer grazing and to make hay for winter feed. The Normans were not farmers. Common land around Saxon settlements became the norm for the next thousand years, until enclosure took its place.

Pasture to feed draft cattle and sheep was most important in medieval times throughout northern temperate areas farmed by man, bearing in mind that all cultivations were done by oxen. Where agricultural communities existed, trees were cleared to make way for grassland supporting herds of grazing livestock. Permanent pasture became common throughout Europe, and this 'stability' in land use allowed a permanent meadow flora to develop under relatively stable management conditions. Hay meadows also became commonplace, allowing a special and specific flora to develop. But by the time we reach the latter half of the sixteenth century, there were many wealthy enclosed properties, far more productive than the common land, which was generally unproductively managed. By now man was the main influence on the land and it would be his influence through agriculture that would control the fate of the countryside.

THE FLOWERY MEDE

The appealing concept of a flowery mede (or mead) goes back a long way. In an essay entitled 'The enamelled mead: history and practice of exotic perennials grown in grassy swards', Jan Woudstra of Sheffield University concludes that 'the practice of growing ornamental plants in grassy swards is a human phenomenon, which can be verified with documentary evidence from at least the Middle Ages to the present day'. However, the medieval garden of delight – not of usefulness – set in the forest with, at its centre, a 'flowery mede' was often translated into words and pictures entirely from the imagination. To quote Christopher Thacker in *The History of Gardens*, it was 'a space of meadow grass sprinkled with innumerable flowers, where Adam and Eve, Virgin and Child, knights and ladies will gather for moments of recreation and delight'. There is no suggestion of work done by way of upkeep.

I remember being greatly impressed at the Cluny museum in Paris, in the summer of 1939, by the series of flowery tapestries hanging there, depicting *La Dame à la Licorne* (The Lady with the Unicorn) and *Scènes de la Vie Seigneuriale* (Scenes of Aristocratic Life). They were woven in the southern Netherlands in the 1480s. The plants are stylized, but many of those scattered in the grassy sward and in the millefleurs backgrounds were recognizable, including bluebells, violas and daffodils. Like the flora in the *Hunt of the Unicorn* tapestries, woven some fifteen years later, which hang in the Cloisters of the Metropolitan Museum of Art, New York, they seem to represent a collective memory of the countryside. The latter tapestries have been examined in minute detail, with close on a hundred plants identified. Among them are some of our best-loved meadow flowers – English bluebells, aquilegias, dandelions, corn marigolds, ox-eye daisies, love-in-a-mist (*Nigella arvensis*) and daffodils – as well as the violet for its association with the Virgin Mary.

In the Renaissance, notably in Italy, the new humanist age looked back to classical literature, to connections with nature and a deep love of the countryside as expressed by such writers as Cato, Horace, Virgil and Pliny the Younger. In his letters written between AD 97 and 107, Pliny describes the garden of his country villa at Tifernum Tiberinum:

> My house is on the lower slopes of a hill . . . it faces mainly south, and so from midday seems to invite the sun into the colonnade . . . In front of the colonnade is a terrace laid out with box hedges . . . outside is a meadow, as well worth seeing for its natural beauty as the formal garden.

Renaissance artists, captivated by the concept of the garden in the landscape, romanticized it, as evidenced in their paintings. The Unicorn tapestries help us to identify some of the flowers in *La Primavera*, which Sandro Botticelli (1445–1510) painted around 1478. Here at Dixter, my mother called her first area of planted meadow her Botticelli garden (translated as Bottled Cherry garden by a small boy staying with us). *La Primavera* was her inspiration (my parents visited Florence in the early years of the last century). Venus, with Cupid hovering above her among fruiting trees, welcomes the approach of Spring, who is

In April the bath-shaped basin of the drained upper moat is like a carpet laid out before light-footed mythical figures. My mother raised the *Fritillaria meleagris* – both purple and albino forms – from seed and was apt to plant them out on the moat banks, rather than at the bottom, where she thought it would be too wet, but the plants thought otherwise and have spread down there.

The flower-studded grass in Botticelli's *La Primavera* of 1478 inspired my mother to plant snakeshead fritillaries and primroses in the turf at Dixter.

heralded by Mercury, the whole scene notably filled in with masses of different flowers.

Another contemporary work, this time of literature, was attributed to the monk Francesco Colonna (1433–1527). He wrote his *Hypnerotomachia Poliphili* in 1499, which was part translated into English in 1592 as *The Strife of Love in a Dreame*. His flowery mede, one of the settings for the dream-vision of the hero, is fantasy pure and simple, of course, but the plants are real enough (my interpretations in square brackets):

> the greens thereof, powdered with such variety of sundry sorted colours, and diverse fashioned flowers as yellow Crowsfoot or golden Knop [possibly *Ranunculus auricomus*], Oxeye, Satrion [male orchid, *Orchis mascula*], Dogges stone [*Orchis*], the lesser Centaurie [*Centaurea nigra*], Mellilot [*Melilotus*], Saxifrage, Cowslops [cowslip, *Primula veris*], Ladies fingers [*Lotus corniculatus*]. wilde Chervile, or sheperdes fruites [possibly *Scandix pectenveneris*], wilde Columbindes [*Aquilegia*], Agnus Castus [the aromatic shrub *Vitex agnus-castus*], Millfoyle, Yarrow [*Achillea millefolium*], wherewith Achilles did heal Telephus, and the rust of the same speares head that hurt him. With the white Muscarioli [*Muscari armeniacum*], bee flowers [*Ophrys apifera*] and Panenentes in so beautiful and pleasant manner, that they did greatly comfort mee (having lost my selfe) but even with the looking uppon them.

'Cowslops' for cowslips, by the way, is said to derive from the fact that the seeds germinate in cowpats.

In terms of garden history, however, the flower-studded orchard goes back even further – according to Penelope Hobhouse, in *Gardens of Persia*, to the sixth century BC and the garden of Cyrus the Great at Pasargadae, the remnants of which can still be seen in Iran. The Achaemenid king's 'paradise' garden included 'square plots between irrigation canals, where orchard trees – pomegranates, wild sour cherries, almonds – were underplanted with clover sparkling with tulips and poppies'. Later, Persian court poets extolled their patrons' palace gardens, with a green tapestry of grass and flowers resembling stars rained down from heaven, using them as metaphor for royal magnificence. By the fourteenth century Persian miniaturists, notably in Herat, and in Samarkand and India a couple of centuries later, were depicting flower-studded orchards and meadows as settings for both real and legendary heroes and heroines. Ibn Arabshah, writing in 1436, describes a meadow outside the city of Samarkand as 'a carpet of emerald, on which are sprinkled diverse gems of hyacinth'.

We need to turn to the West and back to enclosed medieval gardens to fill in this sketch of early meadow history. Expanses of flower-studded grass were probably based on the classical pleasant and natural spot out in the country, the earliest instructions for making a grassy lawn coming in the *De vegetabilibus* of Albertus Magnus (1193–1280) written around 1260, but probably culled from classical Roman manuals. The walled garden in Giovanni Boccaccio's *Decameron* looks forward to the formal gardens of the Renaissance but his band of aristocratic storytellers is also pictured in a meadow, described as 'an unfrequented, almost secret "flowery mede" '. 'Beneath the trees on the plain,' writes Thacker, 'there is a "continuous lawn of tiny blades of grass interspersed with flowers, many of them purple in colour".' The scene 'is still medieval', he adds, 'with its quasi-hidden quality, its flower-strewn grass interspersed with flowers, its encircling trees. The middle of the garden was a lawn "dotted all over with possibly a thousand different kinds of gaily-coloured flowers".'

In the eighteenth century, when the landscape was brought right up to the house, artists depicted an entirely green, unornamented sward, since that was the fashion, but it is to be supposed that, in the absence of herbicides, there were many wildflowers also. Woudstra in fact gives evidence of this provided by Dr W.B. Coyte in 1796. Unploughed pasture contained a high proportion of flowering plants, especially of daisies (which require short turf in which to flourish). Coyte quotes William Curtis from *Flora Londonensis*: 'the enameled meads and dasied carpets by being a little more attended to, may prove useful as well as beautiful.'

With the possibility of a neat and tidy lawn, enabled by the introduction of the mowing machine in the 1830s, the return of formality to the garden was easy. The great bedding-out era followed, using new and often tender plant introductions.

REBELLION AGAINST FORMALITY

William Robinson (1838–1935) was the chief name associated with the return to nature and the rebellion against bedding-out, prompted by the wave of new plant introductions. He decried the monotony of the smooth, weedless lawn, writing: 'Who would not rather see the waving grass with countless flowers than a close surface without blossom?'

Robinson was synthesizing a great deal of work and evidence provided by other nineteenth-century gardeners before he came on the scene. Nevertheless, *The Wild Garden* (1870) was a revolutionary work in which he suggested that 'vast numbers of beautiful hardy plants from other countries . . . might be naturalized . . . in many situations in our plantations fields and woods'. He was a great self-publicist and gave credit to others only when forced to. In 1883 his most important work *The English Flower Garden* was published and went through many editions, the introductory chapters changing with the years. My father had a great interest in topiary, which Robinson loathed, describing it as 'the cramming of Chinese feet into impossible shoes'. In later editions of his book he added a snide reference to its use in Northiam, for it was also notable at Brickwall, a house at the other end of our village.

Robinson opposed any type of formal gardening, but there was quite a lot of formality in his own garden at Gravetye Manor in Sussex, including conventional straight-edged beds on terraces round the house. But there was also what he called the wild garden, which had open meadow areas as well as others in the shade of trees. This seems logical enough, but it is sometimes difficult to distinguish which of the plants he mentions are suited to an open, meadow position and which will grow more successfully in partial tree shade.

Never mind what Robinson thought about formality and topiary, *The English Flower Garden* was my mother's bible and her taste for wild gardening was probably derived from him. Our battered edition is dated 1905, the year my parents married. How to plant is described: 'Ten years ago,' Robinson writes, 'I planted many thousands of Narcissi in the grass, never doubting that I should succeed with them, but not expecting I should succeed nearly so well . . . The flowers are large and handsome, and

An engraving from my parents' 1905 edition of William Robinson's *The English Flower Garden* showing how daffodils and narcissus should be naturalized leaving 'a large breadth quite free of flowers'. This reference work was immensely popular and went into many editions.

in most cases have not diminished in size.' This reads strangely now, but it must be remembered that daffodils and narcissi in his day were, at their largest, not so very large. Nowadays such whoppers have been bred that they give to meadow areas, when planted in them, a most unpleasantly gross appearance.

There are wise words on the planting of narcissi in grass:

> There is nothing we have to fear in this charming work save the common sin – overdoing. To scatter Narcissi equally over the grass everywhere is to destroy all chance of repose, of relief and of seeing them in the ways in which they often arrange themselves. . . . There are hints to be gathered in the way wild plants arrange themselves and even in the sky. Often a small cloud passing in the sky will give a very good form for a group, and be instructive even in being closer and more solid towards the centre, as groups of Narcissi in the grass should often be . . . It is always as well to keep open turf here and there among the groups, and in a lawn or a meadow we should leave a large breadth quite free of flowers.

Hear! hear! But I think this may have been written by one of Robinson's many assistants.

Robinson popularized the concepts of naturalizing which were then elaborated by others. But simultaneous to Robinson's publication, the German Hermann Jäger (1815–90) published his handbook of garden art (*Lehrbuch der Gartenkunst*, 1877) in which he dedicated a section to naturalized flowers in woods and meadows. He noted that 'the only rule' in achieving a flowering meadow was 'to copy nature'. It was not the intention just to increase native plants, but also to add exotic plants in the vicinity of walks and seats. Interestingly, he points out that in meadows with long grass it was the aim to add some red and blue flowers, since native meadows contained mainly white and yellow flowers. Such has been my own experience.

AGRICULTURE, AGE AND SITUATION

In the twentieth century vast tracts of grassland were swallowed up as human populations increased, towns and cities spread into the countryside and land was needed for roads and buildings, although species-rich meadows remained in areas not suitable for cultivation, such as rocky Alpine regions. In Britain nearly all permanent pasture disappeared during and soon after the Second World War, when ploughing grants encouraged the cultivation of every possible site that a plough could visit, regardless of whether it was suitable as arable land. Many meadows (as much as 70 per cent of lowland meadows in Britain) were lost to this post-war agricultural revolution, when farmers increased cereal yields and diversified into such crops as oilseed rape (canola), now the fourth most widely grown crop in the United Kingdom. With the introduction of herbicides

RIGHT, TOP Large areas of grassland have succumbed to intensive agriculture, and the use of herbicides and fertilizers along with a change of management has led to a decline of flower meadows. Only where difficult terrain has stood in the way of agricultural progress, as in the Bavarian Alps, have these precious communities remained intact.

RIGHT, BOTTOM By Lake Balaton in Hungary, in June, there is a rich flora where a ploughed field has given way to weeds of cultivation, scarlet poppies and cornflowers predominating.

the cornfield weeds quickly disappeared. High-nitrogen chemical fertilizers that encourage the growth of coarse plants compounded the problem, wiping out wildflowers along with the broad-leaved plants that were considered the enemy of crops throughout the developing world. Elsewhere, in places like South Africa and Hungary where agriculture is less intensive, arable crops still support a wonderful annual flora.

Intensive agriculture, then, is the enemy of wildlife, be it animal or vegetable. To get a more balanced view of meadow communities and, as it were, go back in time to a pre-herbicide era, I visited Hungary which, owing to its central position in the European landmass, has a very rich flora. Plants (and fauna) of interest can be found anywhere you choose to stop (aggressive drivers stay away, please). Like all the countries that used to comprise eastern Europe, as also Turkey, Hungary is good for meadows because its farming economy is so poor that herbicides and fertilizers cannot be afforded. As a result, grassland has not been 'improved'. Hungary has a total population of only ten million and, as in many lands, there has been a tendency for migration from the country and into the cities. There is a great deal of meadow area but, as everywhere else, it is fragile and liable to degradation, easily overtaken, especially in the wetter areas, by trees and shrubs.

The Hortobágy National Park in Hungary.
LEFT, TOP In June this great tract of flattish land, open to big skies, has bands of colour created by achillea, fescues and other flora that reflect changes in the soil's moisture and alkalinity.
LEFT, BOTTOM A mat-forming thyme, *Thymus pannonicus*, with small copper butterflies supping its nectar.
BELOW At closer range, plenty of interresting plants are seen to contribute. Here are red fescues with *Artemisia pontica*, a wormwood with lacy grey foliage that makes a forest of upright shoots.

I first visited the country more than twenty years ago but on this occasion it was at the turn of May and June and there was a drought on, the vegetation often crunchy underfoot. A curator at Vacratot Botanic Garden (some 25km/15 miles from Budapest), Geza Kosa, was my – I should say our, since I was with Fergus Garrett and his partner, Amanda Ferguson – most helpful host. He was able to take us to many of the areas best suited to our purpose and he had access to the Nature Reserves which are not open to the general public.

The Hortobágy National Park, without tracks, on flat land (though bumpy) and extending as far as the eye could see, is, in the main, dry and alkaline but streaked in different colours, according to changes in the vegetation (themselves largely caused by varying alkalinity). Sometimes green, sometimes overlaid or interleaved with warm brown (caused by the dwarf *Festuca pseudovina*), under a blue sky broken by fair-weather cumulus clouds, it was a great sight. Neither were we disappointed when we looked down at our feet. This was just one site where we saw a lot of the dry-soil-loving *Limonium gmelinii*, which we would call sea lavender. It was the usual mauve, I understand, but late-summer-flowering. There was dropwort, *Filipendula vulgaris*, which we associate with chalk downland; a parasitic *Melampyrum*, *M. arvense*, which as field cow wheat was formerly widespread in England as a weed of arable fields on dry and chalky soils; and a wormwood, *Artemisia pontica*, with lacy grey foliage. Then, almost prostrate, a colony of *Rosa gallica* in full bloom; single, clear deep pink flowers shading to white at the centre, and it was scented. So was the close-by *Thymus pannonicus* – mat-forming, purple and exceedingly popular with small copper butterflies.

In autumn, we were told, the landscape becomes blue and yellow with *Galatella* (syn. *Aster*) *sedifolius* and *G. linosyris*, which has no rays, just tubular florets. In every meadow area the scene is made fascinating by these constant changes of content and through the seasons.

Meadows developed in the first place because the great Hungarian plains, once wooded, were deforested and marshland was drained to provide grazing for some two hundred thousand cattle in small herds. With evaporation, the soil became much more saline, alkaline and impoverished, leading to the development of typical shortgrass steppe with an exceptionally rich flora and fauna. Now the number of cattle has been reduced to only two thousand, but the herds are larger and their grazing is controlled, some animals being confined to paddocks. Large-scale farming and market gardening is concentrated near to the urban centres. This intensification of farming practices has meant that vast areas previously grazed are not being touched, allowing tress and shrubs to move in. The future is dicey: unless they are managed by cutting, burning or grazing, the grassland flora will be lost. Unless support can be raised to service, let alone develop them, they seem likely gradually to return to tree cover.

The meadows that remain and are still operative, mostly for hay-making (generally cut twice, in May and in July), offer a great range of habitats, which change quickly within a very small distance, according to whether the soil is wet or dry, rich or poor, open or shady, disturbed or undisturbed. Some plants take advantage of all situations, while others thrive in specific ones. In both wet and dry we were much taken by a 2m/7ft-tall umbellifer, sulphurwort, *Peucedanum officinale*, with finely divided, glaucous foliage. Some of the most abundant and colourful flora was along roadsides, where the ground, for one reason or another, had been disturbed. Plants are great opportunists. Whereas an established meadow sward might be too dense for some of them, the dry, open situation presented by a disturbed verge can provide just the conditions they need. (In some cases seed lies dormant for years, to germinate when exposed to light.) One of the most dazzling sights was the purple of *Salvia nemorosa*, aromatic in quantity

Plants we saw in Hungarian meadows. **CLOCKWISE FROM TOP** *Salvia nemorosa*, with spikes of purple flowers; a spurge, *Euphorbia seguieriana*, one of the commonest flowering plants of central Europe; and tufted vetch, *Vicia cracca*, which has a huge European distribution; *Asclepias syriaca*, from eastern North America, is a coarse but telling alien ingredient in Hungarian meadows, spreading aggressively on cleared ground where the vegetation is still quite thin.

and in sun-warmed air in a way that you are unlikely to meet in an English garden. There was the equally showy *Lathyrus tuberosus*, with carmine red pea flowers.

It was common to see large colonies of *Euphorbia seguieriana* (30cm/1ft). I cannot but look at plants, wherever they may be, with a gardener's eye. This is a tremendously variable species and only occasionally do you see it looking really effective and brightly lime-green-coloured. Near to the best colony, on a slope too steep to cut for hay, as the rest had been, *Gypsophila paniculata* was in bud. *Phlomis tuberosa* (90cm/3ft) was just out – a branching perennial with half a dozen whorls of hooded, mauve, labiate flowers on each stem. In all, a handsome aggregate.

Typical of the lush vegetation in a wet meadow we saw comfrey, *Symphytum officinale* (popular with bees), whose colouring and flower size varies greatly; tansy, *Tanacetum vulgare*; the fox sedge, *Carex vulpina*; and a 1.2m/4ft-tall fescue, *Festuca arundinacea*. A great show was being made by *Melampyrum barbatum*, a semi-parasite of similar behaviour to yellow rattle, also of the *Scrophulariaceae*, as most of them are, but a haze of mauvy-pink. Legumes were represented by *Coronilla varia* and *Vicia cracca* (tufted vetch). We saw these in a verge elsewhere in dramatic abundance, side by side. The coronilla's whorled heads of pinky-mauve flowers were being fed on by blue butterflies. In quantity, it was strongly fragrant on the air. Dry, previous year's remains of *Carlina vulgaris*, the carline thistle, lover of alkaline conditions, were there. We frequently came across *Asparagus officinalis*, our edible asparagus, which is a sparsely represented native in England where it grows on coastal sand dunes.

Stipa borysthenica, a grass which bends to the wind, its awns being as much as a 30cm/1ft long, was represented but much more abundantly and dramatically in another meadow area, dry and sandy, on a site that was last ploughed fifty years ago. In drifts and backlit by the sun, it looked fantastic in dazzling silvery white. (*S. pennata* and *S. barbata* are similar species that we grow in our gardens, but they are always short-lived, it seems.)

Several species of *Stipa* are among the most arresting grasses of the Hungarian meadows. Here is *S. borysthenica*, its long silky inflorescences blown horizontal in the wind. When its seed is ripe, the seed heads spiral into the ground, thereby increasing their range.

The quaking grass, *Briza media*, was plentiful and charming as ever. By and large, however, there was such an arresting tapestry of other vegetation that one was apt to overlook the grasses.

Here, in this dry area, bushes provided just the cover and breeding ground they needed to nightingales. I have little doubt that nightjars would have appeared in plenty at dusk. I remember them from my earlier visit. The meadow had been slightly invaded by shrubs (the woodies are always ready to move in, which is what makes the meadows so fragile), but not seriously. It was incredibly rich in species and it was hard to tear oneself away – so much so that here I pick out for mention only those that don't come in Chapter 7. We saw a good deep 'blue' form of *Salvia pratensis*; I wish I could get hold of that. Its colouring is so often weak and muddy. The musk thistle, *Carduus nutans*, is a handsome brute, its flower heads proudly presented. There was a clumpy, 35cm/14in-tall clover, *Trifolium medium*, with white flower heads that stood out distinctively. The pale yellow mignonette or weld, *Reseda lutea*, makes a tall (90cm/3ft plus), well-structured plant. It is a biennial generally found on disturbed, alkaline ground, in England. *Achillea ochroleuca*, of dry, sandy areas, grows only 30cm/1ft tall at most and, although not yet out, would clearly make a good display in quantity. It is yellowish white. The *Dianthus*, which I originally mistook for *D. carthusianorum*, with domed corymbs of magenta flowers on long stems, was actually the smaller-growing and less intensely coloured *D. giganteiformis* subsp. *pontederae*, and looked pretty good when massed. *Centaurea sadleriana* (designated an endangered species) is bold, tall (1.2m/4ft) and larger than our greater knapweed, *C. scabiosa*, its leaves deeply cut.

There were damp areas and some of these were being taken over by what our host described as a 'bloody American weed', the goldenrod, *Solidago canadensis*. A lot of *Colchicum autumnale* here, of which we saw only the foliage, of course. The handsomest ingredient, head and shoulders above the surrounding herbs, was *Veratrum album* (1.2m/4ft), often in clumps, in full sunshine and just coming into flower.

On the other side of a small, fast-flowing river that formed one of the boundaries, there was a really wet meadow, a protected field cut once a year but never grazed. Here everything was lush but not colourful: 2.5m/8ft-tall reeds, *Phragmites australis*; common yellow loosestrife, *Lysimachia vulgaris*; valerian, *Valeriana officinalis*; colchicums; a white bedstraw, *Galium palustre*; the variable grass (tall here) *Deschampsia cespitosa*; comfrey, *Symphytum officinale*; and a tall grass, *Molinia hungarica*.

Unfortunately, like elsewhere in the developing world, Hungarian agriculture is 'improving' as larger farms and richer farmers use the land more intensively. This development puts increasing pressure on precious meadow land. The good work of the Hungarian conservationists will, let's hope, offer some sort of resistance to this change in the most significant areas.

The stately perennial *Veratrum album*, with big paniculate inflorescences and ribbed leaves that are no less striking, is one of the boldest and most visible of meadow species in parts of Hungary. It relishes the openness of a meadow whereas the deep-maroon-flowered *V. nigrum* prefers some shade.

2. A Gardener's Meadows

The underlying principles of meadow gardening are the same wherever you live. Having sketched the background story, I am starting with what I know pretty intimately, on home ground, and working outwards from there. After all, interest in the meadows at Dixter is my main reason for wanting to write a book on the subject in the first place. Here at Dixter, I have learnt quite a bit about meadows and a record of my personal experiences must have some value for those who are likewise interested, particularly gardeners seeking inspiration for their own. Like anyone else's, our meadow areas are a dynamic, ever-changing tapestry. They are a huge field for experiment, of which I should like to do much more, for I am ever conscious of the limitations of what I have already achieved. But still, the developing picture is encouraging and gives me much joy, as equally it does to my great confederate, Fergus Garrett.

As I have said, when I was young, conservation was scarcely talked about. Our aims at Dixter were totally different. We wanted to create a flowering tapestry of as many colours as possible, with green the overriding leitmotif. That is still our aim, which is the reason for our introducing alien plants, like the North American *Camassia*.

With the Dixter meadows having, above all, an aesthetic purpose, I am free to cut

Flanking the front path at Dixter are two areas of meadow enclosed by formal yew hedges. They were flowering lawns from the time the garden was laid out; it was the way my mother, Daisy Lloyd, wanted them and I like it. Late April–May is the season of the North American native *Camassia quamash*. This not only makes clumps but also self-sows, which is what naturalizing is all about. I introduced these bulbs, a hundred at a time, when they were cheap to buy. They no longer are.

them when it best suits me (and the meadow) and I can add or remove plants at will, introducing non-native plants if that will be to good effect. (I am mindful of ecological factors but eschew narrow-minded ecological claptrap.) I don't have to concern myself with conservation issues and I am not under pressure to make the land earn, as are the owners of agricultural land. Nor am I bound to manage the meadows cost-effectively, such as besets those in charge of public parks, although Fergus and I are well aware that meadows are by no means the very low-maintenance option that some people would have one believe. That said, the disappearance of nearly all permanent pasture since the 1940s means that, as some of the last pockets of undisturbed grassland, our meadows are areas of conservation as well as being an inspiration to other gardeners. They are certainly havens for wildlife.

When my parents bought Dixter (the Great was added later by my father to distinguish us from Little Dixter, next door) in 1910, it had been on the agents' books for ten years and was semi-derelict. There was no garden, but there were areas of meadow around it. They existed beneath trees in two orchards of mixed apples, pears and plums – the trees planted along ridges which show that the land had been ploughed in medieval times. There was meadow in the once-quarried area around the horse pond (where farm horses were led in to drink) and there was the bath-shaped piece of moat where my mother made her Botticelli garden, mentioned in Chapter 1. There was rough grass in a hollow next the house, where there had been a sewage pond, which was drained and planted up with lilacs, which are still its principal theme, though the turf is important too. Most importantly, perhaps, at any rate most conspicuously, there were two areas of rough grass, one either side of the straight front path leading from entrance gate to house porch.

After the garden had been made, other residual bits of meadow remained as areas of grassland, grazed by sheep. The sheep have long gone and we now have more than half a dozen areas of perennial meadow within the garden, including my experimental prairie, made with plants raised from seed collected in Minnesota, which I discuss in Chapter 10. Now I give you a brief historical overview of our main meadows, with a few of their current seasonal highlights.

THE ORCHARD MEADOWS

The orchard meadows are on gently sloping land running down to the lower moat. The lower orchard, in my parents' day, was never truly a part of the garden, but the upper orchard most certainly was. It was planted up as soon as the garden was made, around 1912, as a daffodil feature over one third of the area, with crocuses and snowdrops preceding them in the fine turf between. The different kinds of daffodil were skilfully grouped: yellow alternated with white; early, mid-season and late were suitably separated with wide margins of green sward, influenced no doubt by Robinson's advice in *The English Flower Garden*. In only one area was there a heterogeneous mixture. They flower in succession from late March (*Narcissus* 'Princeps') to May (*N. poeticus* var. *recurvus*). We had (and still have) a daffodil Sunday every April, when the village was admitted free and my mother liked to give everyone a bunch, in the picking of which my sister (only eighteen months older than me) and I were heavily involved.

My mother had strict rules about the dead-heading of daffodils into which chore (not a word then in use) we were also impressed. It would have been most immoral simply to have collected the dead heads, leaving the green stems (though it might, marginally, have been better for the bulbs). The whole stem had to be picked and we had to count how many we had picked, being paid a penny a hundred for them. The habit of counting became ingrained, and I have never got over it.

After the daffodil leaves had withered, the grass was cut, by scything, and carried as hay. That was in June. Subsequently, the whole orchard was netted in with sheep netting and sheep grazed it until the autumn, so no further cuts were needed, but the sheep's droppings returned fertility to the soil, which is basically undesirable. The sheep treatment and enclosure with wire netting has long since been given up and grass-cutting machinery has made management far simpler.

The daffodil orchard was planted around 1910–12 with varieties fashionable at that time, arranged in groups of the same variety, in this case the trumpet daffodil 'Emperor'. They do not always flower as freely as this – much depends on how the weather suited them the year before.

LEFT Looking across the lower moat towards the house, in spring, there is a tapestry of wildflowers on both sides, all of them self-appointed. Wood anemones, *Anemone nemorosa*, predominate, with clumps of primroses sprinkled among them.
BELOW In a part of the orchard that was once cultivated, and interspersed with damson trees, the turf is coarser than elsewhere, which suits cow parsley, alias Queen Anne's lace, *Anthriscus sylvestris*. Here it is making a May show with buttercups and the late-flowering poet's narcissus, *N. poeticus* var. *recurvus*, which is sweetly scented on the air. The emerging foliage of the tall yellow daisy *Inula magnifica* (illustrated on page 131), can be seen on the left.

In May there is a great surge of growth engulfing the fading foliage of the daffodils and supporting clumps of the poet's or pheasant's eye narcissus, which has the best fragrance of all. One of the most exciting features of the orchard meadow is the huge quantity of orchids, some of which have naturalized since my mother introduced them in the 1930s. The brightest addition we have made to this area is *Gladiolus communis* subsp. *byzantinus*, which makes dazzling spots of magenta in late May and early June.

I love the orchard in July as much as at any time. There is a haze of dusky pink from late-flowering common bent (*Agrostis capillaris*) and, mid-morning, a great display from autumn hawkbit, *Leontodon autumnalis*, whose double yellow daisies are quite a light shade. The flowers turn to the sun, which is why I enjoy standing near the top of the long border, with the sun on my back. The tufted vetch keeps on flowering, as does the lesser knapweed, *Centaurea nigra*. The life cycle of most invertebrates is completed by the end of August (common blue butterflies are

notable in the latter part of the season), so we hold off grass cutting till then. There are plenty of jobs to get on with. The orchard is given a second cut in late autumn in readiness for a knockout display of crocuses from January to March.

The lower part of this orchard is at its brightest in spring, with carpets of wood anemones mixed with clumps of wild primroses. Nearby, in what was formerly a damson grove, there is a concentration of poet's narcissus in the rough grass. To make solo features in July, we grow a very widely spaced colony of *Inula magnifica*. We leave their 2m/7ft skeletons till late winter, though the dead leaves do hang rather like the garments of a scarecrow. They get a mulch when they are cut down.

THE UPPER MOAT

Being close to the house, where my parents did not want water, the upper moat was drained and turned into a fresh piece of meadow. You have a good view of it from the terrace above, where there is a low, raised wall, just the right height for sitting and looking. Because the area was originally a moat, which collected dead leaves and other nitrogen-rich contents, the grass grew more strongly and coarsely than elsewhere and does to this day, nearly a century later, necessitating three annual cuts, rather than our normal two. Summer snowflakes have flourished in the dense turf ever since my mother planted a little colony here in her Botticelli garden. First, though, in the first half of March, the Dutch crocus hybrids are at their best. They have self-sown freely over the years and are now so thick that they make my heart bounce, given a sunny spring day. Narcissus and snakeshead fritillary, which my mother planted on the moat banks, have spread to the bottom, where she thought it would be too wet, but they thought otherwise. The summer snowflakes reach their peak after the crocuses have faded, when the wild celandines, which have come of their own accord, are flowering all over. The polyanthus primroses my mother introduced at an early stage still hold their own against all comers.

June to July displays are provided by meadowsweet, *Filipendula ulmaria*, and a rich-purple-flowered strain of English iris, *Iris latifolia*, which is bulbous and thrives.

OPPOSITE, TOP The upper moat in April: the terrace wall behind, covered with mauve aubrieta, provides a cheerful background. Celandines have come of their own accord and open wide to the sun. The summer snowflake (*Leucojum aestivum*) has naturalized in the dense turf and flourished ever since my mother planted a little colony here in her Botticelli garden. Snakeshead fritillary (*Fritillaria meleagris*), which she planted on the moat banks, has spread to the bottom.

OPPOSITE, BOTTOM A great wave of red valerian (*Centranthus ruber*) follows the aubrieta in the walls overlooking the upper moat, which, in May, is bright with buttercups and *Gladiolus communis* subsp. *byzantinus*.

THE FRONT MEADOWS

The two areas of rough grass on either side of the front path were planted around by my father with yew hedging so that it looked as though they should be tamed and mown, but they never were (that was my mother's influence). To this day, in summer, before the first grass cutting in August, people get as far as the front gate, lean over it, hoping to see as much garden as they can without entering or paying, and are amazed to be confronted by what to all intents and purposes appears to be a hayfield. Sometimes we are pityingly sympathized with for labour being so dear these days. We smile and carry on as we always have.

The meadow areas may at times look dishevelled and disorganized but this is far from being the actual case. In fact, I love the summer season as the grasses grow taller; they bend before the wind and the light catches and reflects their stems. Buttercups and red clover sparkle among them in May, and sorrel makes a haze of red in June. To the tidy-minded, the front meadow is more appealing in March and April, when Dutch crocus hybrids (*C. vernus*) coincide with Lent lilies (*Narcissus pseudonarcissus*) to make carpets of colour in the short turf. As the grass grows taller, it supports blue and white camassias, which gleam among the green most effectively. By July, the camassias' seeds are ripening and we make sure they are shed before the power scythe moves in the following month.

To curb the length and strength of grasses, we have established the semi-parasitic hay rattle (yellow-flowered in June), *Rhinanthus minor*. It is pretty in itself but annual and the seeds soon lose their vitality. To establish a new colony, they should be gathered in early July, as soon as ripe, and scattered on the turf surface by September. They will germinate in February and be self-perpetuating from then on. There is now no problem from tall grass flopping over the front path.

This photograph taken in the front meadows shows the trunk of one extant walnut, the only survivor of five. The soil is shallow and does not suit them. But it does suit a wide range of meadow flowers. Lent lilies, *Narcissus pseudonarcissus*, which Daisy Lloyd raised from local seed and planted out, and crocus are prominent in spring when the turf is short.

THE TOPIARY LAWN

I have only allowed our topiary lawn to become a meadow within the last ten years (at the suggestion, I may say, of a young friend, Graham Harvey). My father was a keen golfer and this was his putting green, where not a single weed was allowed to intrude. My mother and I ensured this by creeping about on it with kneeling mat and daisy fork. But my father died in 1933 and no one was subsequently interested in its golfing associations. Apart from the occasional application of a selective broadleaf weedkiller, I neglected this area, which, being badly drained, became invaded by sedges, dried out and turned brown in summer and, in general, did me no credit at all. The fact that the ground was poor made it ideal for conversion into a meadow with no anxieties about coarse grasses dominating.

As soon as I allowed the grass to grow long, it picked up in vigour (although never so vigorous as to be a problem). Invading weeds became my friends and it is now most colourful, especially in the mornings, when the various yellow daisies are open to and facing the sun. Notable among these weeds is the autumn hawkbit that I also enjoy in the orchard – it jollies up the scene from June till late summer.

Towards the end of June, the topiary pieces, shaggy with young growth, seem to be swimming in the deep sward where the soft pink flowers of common bent hold sway over white clover, conveying its honeyed fragrance. A month later, this has become a sea of undulating, bleaching grass – a magical sight in contrast to the solid topiary yews.

I have not planted much into the grass and no bulbs, so our regime to date has been to cut it like a lawn in February; then not again till September, when we need to reach the topiary yew specimens for their annual cut. With the long grass shorn and the topiary pieces smartened up, formality is at last restored. We can enjoy the long winter shadows cast by the crisp topiary shapes and look forward to the wild orchids. Two species, the green-winged and the spotted, have come into this area without encouragement (their seed is tiny and very light and is blown in by the wind); we have even recently discovered a plant of the southern marsh orchid. Their arrival has not affected our cutting regime here, though it does prevent me from fitting in a second spring cut (as John Sales does in his Gloucestershire garden, even as late as May, his mower blades set high to avoid the orchids). But by the end of March the self-sown spotted orchids are already taller than the grass itself, while the green-wingeds will be flowering in some years by mid-April.

Looking across the drained upper moat, with *Gladiolus communis* subsp. *byzantinus* in the foreground. There used to be beautifully manicured lawn beyond the path. Here my father practised putting, for he was a keen golf player. For the past ten years it has become meadow.

THE HORSE POND

Being, basically, subsoil, the area around the horse pond was very starved when my parents came to Dixter. My father planted it up with heather (ling, *Calluna vulgaris*) and whins, as he called gorse, both reminding him of the Scotland in which he was brought up. His mother and maternal relations were all Scots. After his death (when I was twelve) in 1933, the horse pond area became very rough and infested with brambles and rabbits. Oaks sowed themselves into it from across the entrance road, the forstal (a local word). A few have been left and are now sizeable trees themselves. During the Second World War, the area was fenced and geese and chickens were kept here. Also a house cow or two (Jerseys). A land girl, our neighbour's daughter, looked after them. Geese make a horrible mess with their droppings, so there was no great joy from the area as a part of the garden. When this had all come to an end, we cleared away the scrub. The rabbits were controlled by myxomatosis, around 1950, much to our joy and relief, though we had depended on rabbit meat as extra-rationing all through the war years.

So this became enjoyable meadow too, managed with hand tools until relatively recently. Being a natural pond on a clay substrate, the grass banks come right up to the water's margin. Its mowing is mechanized now, but drainage remains an on-going problem. It would seem that no lasting pattern of drainage pipe has yet been devised.

Meadow goes right up to the margins of our horse pond, where farm horses used to be led in to drink. In May moisture-loving blue *Iris sibirica* is in flower among clover and buttercups. On the far side of the pond there are wild yellow flag irises, *I. pseudacorus*, and Bowles's golden sedge, *Carex elata* 'Aurea', which retains a good colour till August.

All become blocked after a few years. We have a particular problem with rushes (two species of *Juncus*) here, where drainage is worst. The root system of rushes is so dense and tangled that other species are unable to compete on equal terms.

This area is particularly popular with visiting members of the public. Water is always an attraction, and very much so in this case where we have a legacy of thousands of fish and, anyway, there's always a lot going on in and around a pond, especially where the ground is not over-shadowed by trees. Without any mowing, a considerable area around here is kept short by feet, after we open to the public. Until April, however, dog's-tooth violets (*Erythronium dens-canis*) have nothing to fear from shoes and boots (except mine). I make a point of visiting the horse pond every pleasant day in early March to catch the sight of the sudden appearance of the dog's-tooth violets as dark, chocolate-coloured clusters of unfurling leaves, with narrow flower buds nestling at the centre. In open-textured turf on the once-quarried bank that is set back from the horse pond, the clumps of *E. dens-canis* alternate with those of *Narcissus minor*, which is just the right diminutive size as a companion, and of primroses, which I had to introduce here, but they are well established now. In a month's time *Erythronium revolutum* is flowering among bluebells in the dappled shade beneath the trees on the bank.

Through April, often in March, the cuckooflower (*Cardamine pratensis*), which I have always known as lady's smock, is at its peak in heavy, wet soil by the pond. This is one of the food plants of the orange tip butterfly, which is such a joy to see on the wing at this time of the year. By May the grass in the damp ground fringing the horse pond is tall and lush. Native flag irises come in a rush of yellow at the turn of May and June. To these I have added other irises and filipendula (queen of the prairie, *F. rubra* 'Venusta', which is soft pink). We wait until September to start cutting this area – there is a very worthwhile late display of lesser knapweed, *Centaurea nigra*.

Goldilocks (*Ranunculus auricomus*), lady's smock (*Cardamine pratensis*), wood anemones (*A. nemorosa*) and cow parsley (*Anthriscus sylvestris*) make a lively little meadow scene in the hollow in spring, contrasting with a bedding scheme in the foreground of a mauve wallflower, *Erysimum linifolium*, and the late orange tulip 'Dillenburg'.

THE HOLLOW

Being damp (surface water from the high garden drains into the hollow), the turf beneath the lilacs in this area is planted with kingcups and clumps of a handsome red-flowered comfrey which I collected from a ditch-side in Romania (Transylvania) some twenty-five years ago. Around the hollow is a lively display of spring goldilocks, lady's smock and wood anemones.

3. A Setting for Wildflowers

Meadows in gardens need careful siting. Dixter's gardens have a lot of bone structure. Lutyens designed the garden walls and how the hedges should run. My father had an eye for strong design and a strong interest in topiary. My mother, who was my great gardening influence and inspiration, was a plantsman – like me, she loved plants for their own sake. For both of us, the architectural setting was, and is, a boon.

A dark background to a sunlit area of meadow is the most flattering, and can be provided by trees or shrubs and, in particular, by evergreens. But the sward itself is best not interrupted by large or solid obstacles. A few widely spaced trees will allow the eye to travel freely, while the grass and its contents will not become unduly shaded. An orchard of standard trees provides an admirable setting, provided the orchardist's good husbandry is nil and does not include fertilizers.

Interruption is best avoided, but the contrast of swaying grasses and sculpted yew in our topiary garden is very effective. Indeed some kind of contrast is needed between a meadow or prairie and the rest of the garden. It conveys a message: 'this meadow is intentional; it is not just a neglected area.' On a small scale, the difference between meadow and manicured lawn may be enough. At Dixter, the margin of our upper orchard runs parallel with the highly organized

Near-wild contrasted with formality. Besides crisp yew, the topiary lawn now has other shrubs. Here the smoke bush, *Cotinus coggygria*, is in bloom, its smoky flower heads looking like a continuation of the flowering grasses beneath.

mixed border, called the long border, with only a 1.5m/5ft-wide flagstone path and a parallel strip of mown grass between them. Some people think it is a shame that a wilderness of long grass should visually 'spoil' the setting for the border, but to us this seems a brilliant juxtaposition of near-wild and tame. The dividing strip of mown grass, which makes the transition, is essential. It accustoms the eye to the change of subject. From a practical point of view, a strip of mown grass prevents the meadow from spreading.

An edging of brick or stone may also be used to contain a meadow or prairie, and has the advantage of no mowing. We keep the flagstone path that bisects our front meadow free of weeds (no volunteer seedlings are allowed to stay in the paving cracks here) to contrast with the masses of flowering plants on either side. Until we reduced their vigour by introducing hay rattle, the grasses fell over the path after heavy rain and this was noticeable because there is no strip of mown grass dividing the path from the tall grass. Luckily the path is wide enough to allow people to pass through, an important point to consider when planning your paths adjacent to or through meadow areas.

ABOVE The circular steps designed by Lutyens, which project into the orchard meadow, are cleared of all 'weeds' in winter, but in summer are engulfed by red and white valerian, *Centranthus ruber*, and wildflowers from the orchard meadow. **OPPOSITE** A path through the orchard meadow at Dixter curves not for the sake of making an artificial wiggle but for a good reason, as when it needs to avoid overhanging branches or the trunk of a tree.

Another essential is to have mown paths through an area of meadow. Not only does this involve you more closely with it, and allow easy access, but paths confirm the message that the meadow is meant to be there, and add: 'this meadow is to explore, but I will direct you.' The positioning of these paths is especially important in public places: they encourage visitors but also discourage them from cutting across the meadow, thus protecting the meadow plants, and wildlife, from disturbance. They also reduce the likelihood of people picking up barbed seeds or unwelcome insects. But there should not be too many mown paths, fussing the area up. They should not wiggle meaninglessly. They should have some relevance to contours and, if they take on a curve, it should be to avoid an obstacle, probably a tree. Sometimes a straight path will seem correct, leading to some feature at its end. Not necessarily a seat or statue or a sundial or fountain. A plant will often be meaningful enough.

At the height of the season, the paths will probably need mowing biweekly. Trampled by the hordes, they may need repairing in patches. At Sissinghurst in Kent, the heavy soil in the orchard prompted the laying of drains – for the sake of the trees, but, writes Tony Lord in *Gardening at Sissinghurst*: 'It was particularly

important to achieve good drainage along the grass paths to help them withstand heavy wear and tear in wet seasons.' Even so, they have to be repaired: the gardeners rotovate and resow the paths with a grass seed mixture designed for race-courses that contains mostly dwarf ryegrasses that remain hardwearing and green throughout the summer.

Visitors of the human kind are to be encouraged, but we have others we try to deflect. Badgers regularly dig holes in our meadows in which to defecate. These turn from fine turf into coarse, second-rate vegetation, largely invaded by hogweed. We use mini-hurdles, made locally from split chestnut, as discouragement.

From the aesthetic point of view, a meadow can be just the thing to add a touch of near-wild to an otherwise too-tame environment. In practical terms, you have to think about your neighbours, particularly in suburban areas. In the United States of America there are so-called weed ordinances banning lawns over a certain height, enacted to prevent abandoned property becoming an eyesore. You may need to convince your neighbours in advance of turning a lawn into flower-studded grass.

It is true that 'weed' seeds do move across from the meadow into our borders, but we have never found this a serious problem. In fact, weed seeds, such as thistles, blown in from neighbouring fields, are more of a nuisance. I welcome seedlings of

LEFT Meadows make an effective contrast with formality and complex planting. A flagstone path and a strip of mown grass separate our highly organized long border from the wildness of the orchard meadow. Weeds sometimes move across from the meadow into the border but are rarely a serious problem.
RIGHT Generally, very little seeding occurs from the garden borders into our meadow areas. Where a yew died in the topiary lawn, a large gap was quickly filled by self-sowing annuals – love-in-a-mist (*Nigella*) and, seen here, the poached egg flower, *Limnanthes douglasii*. The latter spread into the clover and grasses in the meadow, blending the wild with the tame.

the common spotted orchid which turn up (among bearded irises especially) and, like moon daisies (or ox-eye daisies), invade pavement cracks in every part of the garden. They are a link with the wildness of the orchard meadow and Lutyens' formal landscaping in the garden proper. Conversely, very little seeding, if any, occurs from the borders into the meadow areas, which shows how grass-dominated our meadow communities are.

Our climate favours grass growth, and escapees are rarely a nuisance in our managed meadows. This certainly isn't the case in North America, where exotics have caused many problems. In Hungary in a large area that had been cleared, two and half years before, of a crop of poplars and then just left, the vegetation was still quite thin. An outside invader was the coarse American perennial (food plant of the handsome monarch butterfly) *Asclepias syriaca* (see page 67). 'Recently cleared' is telling: the age of a meadow makes a great difference to its content. First to appear are the plants which benefit from the lack of competition.

Another way to contain a meadow is to create a barrier of plants. Nigel Dunnett's work with shrubs that may be used to provide the transition from woodland to meadow or prairie is described under 'Dynamic Woodland Edge' (page 225).

4. Creating Permanent Meadows

It will be appreciated how difficult it is to give detailed advice on how to start or what to plant in a meadow when I say that our own examples vary very considerably over quite small areas. Part of what makes meadow gardens fascinating is that you have to experiment. I include here several experimental meadows we have visited to learn more about meadow establishment. More of these are discussed in subsequent chapters on annual meadows, prairies and in the larger landscape.

OPPOSITE An abundance of blossom and yellow daffodils in our orchard meadow. The dying foliage of daffodils is never unsightly in a meadow as it is swallowed up among the grasses, which make a great surge of growth in May.
BELOW Cow parsley or Queen Anne's lace, *Anthriscus sylvestris*, prefers partial shade and takes over in somewhat enriched soil around a tree. It is no friend to the finer elements in the ideal starved meadow.

Most grassland plants like full exposure to sunlight, plenty of spring rain, and poor soil. The poorer the soil, as a result of low nitrogen, dryness or acid conditions, the easier it is to grow the most varied tapestry of flowers. Once you start to enrich it, for instance by leaving grass mowings *in situ*, certain greedy plants, like cow parsley (otherwise known as Queen Anne's lace), will benefit at

the expense of others, which will be elbowed out. While, in general, the richest communities of plants will be in open situations, there are exceptions to the full-exposure precept. Where there is shade from trees the grass covering will automatically be thinner, even though some grass species are specially adapted to shade. The gaps that shade areas foster allow the opportunity for colonization by shade-loving perennials that would find competition too great in the dense sward of an open site. Such, for instance – needing pretty moist soil – are snowdrops, doronicums (leopard's bane), hellebores and erythroniums (dog's-tooth violets). On the north side of our front-path yew hedge, where there is also shade from a wild pear tree and the turf is a bit thin, snowdrops clump up more happily than in the orchard meadows.

IN TURF

The simplest way to create a meadow is in an existing lawn. Stop using weedkillers and fertilizers, leaving the broadleaved plants to reproduce and introducing wildflowers as seed or plugs. Where the turf (to American readers I should say that this is what you call 'sod') is unduly coarse or weedy, keep it mown very tight for a year. Only at the end of this period should anything be planted. Normal management (Chapter 5) thereafter.

You will get a feeling for the sort of area a plant would like. Thin turf beneath a light tree canopy is ideal for the European dog's-tooth violet, *Erythronium dens-canis*, which goes well in its March season with the small species daffodil *Narcissus minor*.

For the American trout lily, *Erythronium revolutum*, which flowers a whole month later than *E. dens-canis*, also avoid dense turf. This nearly always means that there will be some overhead shade.

If the soil is fertile, it is advisable to skim off the top layer of turf and get to work on the less fertile layers beneath, which also contain fewer weed seeds and will sustain populations of smaller plants. Admittedly, these will take a couple of years to establish, so, unless a catch-crop of colourful annuals is taken, there will be little to impress the sightseer until the second year.

Sowing seed into existing turf is a hit-and-miss affair. In the autumn, cut the turf short, scratch the surface with a rake to remove thatch and leave patches of bare earth. Sow wildflower seed into these patches at 1.5g per square metre/1⁄25 oz per square yard and cover with sieved soil mixed with sand. Lightly rake or roll to ensure the seed makes good contact with the soil. You will need to mow regularly (with the blades set high) during the first year to allow seedlings to establish without being swamped by the existing sward. Planting seedlings and perennials grown in plugs is much more reliable.

Nearly all planting in turf can be most conveniently done with the help of a bulb-planting tool. This takes out a plug of soil 15–18cm/6–7in deep and not more than 5cm/2in in diameter. Avoid the ones that take out a larger plug, as they are quite unnecessarily hard work. If a plant or bulb is more than 5cm/2in across, a few more nibbles with the bulb planter round the edge of the hole will enlarge it. Put a handful of old potting soil at the bottom of each plug hole to give the plant a good start.

I consider the method of planting recommended in our 1905 edition of Robinson to be iniquitous. (The bulb planter may not have been invented at this date, although it came in very soon afterwards. The tools hanging up on the potting-shed door at Calke Abbey include a bulb planter of exactly the same model as ours, and our orchard was planted up around 1912.) This practice, which consists of turning back a sod as a flap, cut on three sides, planting in some bulbs, pretty densely, turning the sod back and treading firmly upon it, still obtains in many public gardens and, I regret to say, is taught to the students at the Royal Botanic Gardens, Kew. It looks what it is: unnatural and altogether terrible.

Plant bulbs into turf according to season but introduce perennials in spring. It makes a great difference to their successful establishment if an area around them is kept free of weeds for the first couple of years. Plant them into a turf- and weed-free circle, its optimum diameter being 45cm/18in. Weeds invading the gap can be spot-treated, early in the year, with glyphosate. If the introduced plant shows any green, protect it with a guard. No further treatment to preserve the weed-free gap is needed, the most important cultural factor now becoming the best date for giving the meadow its main cut (the herbage being removed).

Here it is worth pointing out that if introduced perennials are raised in a peat-based compost, or a compost using peat substitutes (rather than loam), it is essential that the compost be thoroughly soaked before planting; also while the perennials are establishing, after planting. Otherwise the plant roots will never make contact with the surrounding soil, as the peat or substitute shrinks and leaves a gap between itself and the surrounding soil. I am sure that most of my readers have had the experience of a seemingly strong plant, when planted out, simply never taking off. To forestall this, it is a good plan to tease away some, even most (if that will not be damaging the roots) of the peaty compost before planting.

With a spade, cut away an area of turf 45cm/18in in diameter to enable introduced perennials to establish without competition from grass or weeds. In heavy soil the smooth sides of a planting hole made by a spade or a bulb planter may impede drainage and need to be scarified with a fork.

STARTING FROM SCRATCH ON POOR SOIL

Never having had to start from scratch myself (and with no intention to do so), the advice I give here on preparing and sowing a new meadow is largely based on guidelines given in a report by the Weald Meadows Initiative (WMI) on meadow establishment during the first two years, which are the trickiest. (More about the WMI in Chapter 11.)

An alternative to sowing, I'm told (not by the WMI), is to buy wildflower turf from specialist nurseries. I haven't tried it and am not likely to have to. It sounds like a good idea for creating a very small meadow (but why not sow those too?) and a definite, and expensive, cop-out for substantial areas of grassland. Fergus tells me there is another product, called a 'Wildflower Blanket', made from recycled clothing which is shredded and reformed into a felt product. It has its uses for establishing wildflowers on sites that lack a sowing medium and those on which disturbance is not an option. The blanket acts as a seedbed, helps to retain moisture and provides nutrients as it degrades (woollen fibres slowly release nitrogen). On more fertile sites the blanket acts as a weed suppressant. After rolling it out spread a 30mm/1¼in aggregate over it to increase moisture retention and prevent wind and bird damage.

Back to the WMI advice. Prepare the ground by removing existing herbage or 'burn' it off with glyphosate (Roundup) and remove large stones and boulders. From personal experience I would emphasize that, dull though it may be to have to wait, in the long run it is well worth treating an area several times over a growing season, or even two growing seasons, with glyphosate, so as really to cope with those persistent weeds like couch grass and field convolvulus.

Cultivate the soil by surface tilling, then rake over and roll the surface to create a fine tilth for seed sowing in autumn or spring. Autumn sowing allows for the fact that many seeds need a cold spell before germination will occur. Spring sowings often run into drought. Preparing the ground will have created the perfect environment for weeds, so wait a few weeks and either treat them with glysophate or pull them out by hand or hoe.

Deep cultivation is not necessary as this encourages deep rooting and more vigorous growth. You can leave a small proportion of the area compacted to increase the diversity of the vegetation. A fine firm seedbed that is not too deep is ideal. Any imported soil should be low in nutrients. Sand, gravel, subsoils, brick rubble and cinder may be combined in different mixes to produce a variety of conditions. Alkaline waste may also be used to produce calcareous conditions. Allow the area to settle before sowing.

STARTING FROM SCRATCH ON FERTILE SOIL

Rich soil is always a problem when creating a meadow of varied content, especially if the requirement of minimum upkeep is in question. Coarse herbage not only excludes the smaller species but also requires frequent cutting and carrying, which is a prohibitive expense.

One solution is to strip the topsoil to reach the less fertile subsoil, which will be relatively free from a bank of competitive weed seeds. This will sustain populations of smaller plants and, again, you may want to sow catch-crops of colourful annuals to tide you over the couple of years they take to establish. However, it is better, I think, to be strong-minded and not to include annuals in the initial seed mix, whose shading of the nascent perennials in their first season of establishment may not actually kill, but will certainly weaken them.

Pam Lewis opted for scraping off 18cm/7in of rich topsoil – Dorset loam – to make a new meadow at Sticky Wicket, first having considered the effect of altered land levels. She hired a toothed digger bucket (and driver) to do the job. To confirm stripping as a viable solution, I visited meadows in the north of England in Merseyside, at Knowsley and Pickering's Pasture. The former project is run by Landlife (of which more in Chapter 11, where I also include information on the reclamation of Pickering's Pasture).

In mid-July, Richard Scott, one of Landlife's most dynamic officers, took me to see the 2-hectare/5-acre site at Knowsley – a disused farm field, which had been neglected for years and where uncontrolled fires occurred in most summers. A thick layer, perhaps 50cm/20in, of peaty loam had been skimmed and sold as topsoil, which brought in sufficient revenue to fund the purchase of seed with which to colonize the subsoil. This was sandy and easy to cultivate.

If your subsoil is similarly light, you can remove as deep a layer of topsoil as at Knowsley, but a thinner scraping is likely to be enough, say 15–20cm/6–8in, to reach the less fertile layer. On a smaller scale, it is fine to leave a thin scraping of topsoil to get the plants going.

Clay subsoil is much more difficult. Having scraped off topsoil to a depth of 10cm/4in, using a mechanical digger on large areas, thus removing the weed-seed bank, you have various alternatives. If reducing the depth is not an issue, you can reduce the fertility of the subsoil by allowing the herbage to grow and cutting it frequently over several seasons. An alternative to waiting several years,

which also makes good the depth, is to rotovate a fine grade of crushed concrete into the sub-soil as evenly as possible. Crushed concrete is one of the best forms of industrial waste. Brick rubble compacts, if too fine, though it can be mixed with coarser materials. A fine grade (up to 10mm/½in) of crushed concrete is preferable, as it mixes easily with soil. By itself, it has too high a pH for plants to be able to cope. It needs to be mixed with sub-soil in a 50:50 ratio. This will bring the pH down sufficiently to satisfy a calcicole seed mix (the delight of bee orchids), which is then sown on to this at 5g per square metre/⅙oz per square yard, harrowed and rolled.

The scrape at Sticky Wicket in Dorset involved moving 300 to 400 tons of fertile topsoil from a 0.2 hectare/½ acre field to reveal the clay subsoil. This took away coarse grasses, aggressive weeds and much of the weed seed-bank, as well as excess fertility, providing Pam Lewis with a 'clean slate' for her new hay meadow. It also altered levels and affected the way water ran off the area, involving the installation of drains. The spoil was used to create the 'mount'. Seed came from a donor meadow in the area.

CHOOSING SEED

Buy from a reputable firm which can supply seed of local provenance. You can collect local seed yourself (with the landowner's permission), or practise 'hay strewing' (see below). Some seeds may need special treatment before sowing. Select seed ecologically suited to your site – wet or dry, clay, sand or limestone – although whatever you start with, the end product will be the same, those grasses and other plants dominating which are best suited to their environment. Choose common

Hay rattle or yellow rattle, *Rhinanthus minor*, is a semi-parasitic annual that feeds mainly on grasses and is a valuable tool for reducing the vigour of coarse turf. There are other semi-parasitic species to suit your locality. They should be sown within two months of ripening, needing a cold period before germinating.

species that establish easily: the seed mixture is unimportant as long as you avoid coarsely-growing grass species (see Chapter 6). For most soils, choose or make a mix of about 80 per cent slow-growing grasses to 20 per cent wildflowers. Avoid the temptation to increase the proportion of flowers to grasses, as in the long term this will allow greater weed penetration.

The Knowsley seed mix was of non-invasive grasses and a fairly restricted range of British wildflowers, the seed being supplied by Emorsgate Seeds. Flowers that particularly caught my eye were field scabious, *Knautia arvensis*; bird's-foot trefoil, *Lotus corniculatus*; kidney vetch, *Anthyllis vulneraria*; lady's bedstraw, *Galium verum*; greater and lesser knapweeds, *Centaurea scabiosa* and *C. nigra*; common centaury, *Centaurium erythraea*; and autumn hawkbit, *Leontodon autumnalis*. Also, I have to say, the invasive but beautiful, when in flower, Yorkshire fog grass, *Holcus lanatus*, but the vigour of this can be materially reduced, as described earlier, by the introduction of the semi-parasitic hay rattle, *Rhinanthus minor*, and this has been done. Other semi-parasitic incomers are eyebright, *Euphrasia* and red bartsia, *Odontites vernus*, of which more in Chapter 7.

These semi-parasitic plants, not included in the original seed mix, are an important element, increasingly so at Knowsley, where 27 of them (to date) have come in to join the original 16 species sown. Orchids are expected to put in an appearance at any time. Their light seeds get blown great distances and the circumstances are propitious. Stop press: orchids have now appeared on the stripped site, including bee orchids.

One of the greatest population explosions has been by the autumn-flowering devil's-bit scabious, *Succisa pratensis*. Only 10g/⅓ oz of seed, obtained by Scott from a Lincolnshire site, was originally sown, but it is now everywhere, in its thousands. This, if undisturbed, is a notably adaptable species of any sort of grassland, in sun or partial shade. Even so, its habitats have been greatly reduced. It is the sole food plant of the marsh fritillary butterfly, which is consequently under considerable pressure.

SEED SOWING

The soil should be moist at the time of sowing and kept moist for at least two months – you will have to irrigate the area if the rain does not do it for you. Once established, meadows really are low-maintenance from the point of view of watering and will survive with much less than the average needed for beds and borders – as little as 30cm/12in of rainfall per year.

The rate for sowing most native wildflower mixes is approximately 15–20kg per hectare/13–18lb per acre (for small areas 1–2g per square metre/⅛ oz per square yard). Seed can be either hand- or machine-spread on the surface. To help with even broadcasting by hand, mix the seed with damp sand, sawdust or fine soil. Gently rake over to ensure good contact of seed with soil, and roll again. Water thoroughly with a fine spray.

MANAGING A SOWN AREA

Germination is often spread over a year or more. The seedlings must not be overshadowed by weeds or by any heavy top growth. When the young plants are about 7cm/3in tall, cut them back (or graze large areas) to about 4cm/1½in. This will prevent them from flowering and also limit vigorous grasses and many weeds. Remove all cuttings and use for hay or composting. Until the autumn (November in the northern hemisphere) continue to cut regularly, unless your mix includes annuals,

in which case leave cutting again until the autumn. Always remove the clippings.

In the second year the developing perennials will suppress the annual weeds, but mowing should stop between spring and mid-summer. Different cutting regimes will suit different species. A cut in June will favour the early-flowering species such as buttercups, sorrel and cowslips, while a regular August cut will favour the later-flowering species like knapweed, scabious and lady's bedstraw.

The first year after sowing is likely to give a show of pioneer species, such as the aforementioned buttercups and moon daisies, but don't despair. Most meadow perennials take at least two years to flower, lying dormant or spending the first season growing good root systems. That is why seed mixes tend to include some annuals. For the first year you will have to hand weed, or spray out, pernicious weeds, such as docks, nettles and thistles. Annual weeds are not a problem after the first couple of years because mowing prevents them from seeding.

Management once the area is established is outlined in Chapter 5. At Knowsley, however, the grass is cut just once a year, in August, and as its growth is comparatively feeble, it can be chopped up fine and left *in situ* without noticeably affecting fertility. This reduces management costs dramatically.

FAR LEFT A summer cut (in June for us at Dixter) will favour early-flowering species such as common sorrel (*Rumex acetosa*), which makes a delightful red haze.

LEFT *Leucanthemum vulgare* beneath silver beeches at Lady Farm in Somerset. These ox-eye or moon daisies are pioneers in ground that has been newly disturbed but they continue to survive in old meadows, their populations fluctuating, even mysteriously.

HAY STREWING

This is an interesting way of establishing a meadow in the same area as a species-rich donor meadow. The recipient site must have similar geological and geographical characteristics – hence 'the same area'. The hay is cut in early July, baled and transported to the naked area that is to be seeded; then spread very thinly all over that, covering twice the area it came from. The whole procedure needs to be carried out in the same day, the hay still a bit green and therefore retaining most of its seed. It would start fermenting within hours if kept in a bale, hence the need for immediate strewing. The hay acts as a mulch and should be turned regularly or chain-harrowed to knock the seeds into the soil. After two or three weeks, the hay will be dry and should be removed to stop it smothering the seedlings.

You can buy hayfield mixtures, consisting of chaff and seed, but if there's a species-rich meadow in your vicinity, strewing is better for the preservation of local populations and therefore biodiversity itself. Indeed, there are some schools of opinion, such as are held by Quentin Kay, from the School of Biological Sciences, University of Wales, that believe strongly that if seed travels over a significant distance from where it was collected, this works against the preservation of ecotypes.

Donald Macintyre, the owner of Emorsgate Seeds, while acknowledging the importance of preserving ecotypes wherever possible, points out that there has long been considerable movement of seed, first by travelling seedsmen and later with the movement of hay over larger distances as transport in the twentieth century improved. It seems to me that both arguments have a point. In the world we live in we need to be pragmatic if we are to exist. (What is British, when we claim that seed grown is of home origin? Does it only have to have been two years in this country to have become native? There is some, specious, definition).

Macintyre, who lives in attractive, hilly country in Somerset, near to Bath, showed us (on the last day of April) a meadow he had established by strewing hay. Fifteen years previously, it had been a ploughed field; then became a tumbled-down meadow, largely inhabited by the over-bearing, coarse false oat-grass, *Arrhenatherum elatius*, which grows to near 2m/7ft tall. Macintyre grazed it heavily for two years and also weakened it with hay rattle. In the third year, in the second week of July, he hay-strewed it from a nearby field which we could see in the distance, yellow with cowslips. He timed the hay cutting to catch much of the cowslip seed (this plant flowers, and hence ripens, over quite a long period). Sheep are still run on it in August–September.

Macintyre's meadow has a varied flora including, not surprisingly, plenty of cowslip (*Primula veris*); cat's ear, *Hypochaeris radicata* (one of those innumerable little yellow daisies); a hawkbit, *Leontodon hispidus*, the rough hawkbit, which prefers calcareous soils, as here; *Lathyrus pratensis*, a fairly aggressive yellow-flowered pea; and Fergus got a first sighting of the adder's tongue fern, *Ophioglossum vulgatum*, which was quite an excitement.

INTRODUCING PERENNIALS

Perennials can be planted into a seed-sown meadow (the method as described above in established turf or sod), so long as the population of weeds is under control. What you introduce is a field for experiment. Dr James Hitchmough's meadow experiments at Harlow Carr, near Harrogate in Yorkshire, include one in which he is seeking to find out which exotic species will not only survive the competition within a basically native meadow flora, but will increase and also withstand the mowing treatment necessary if it is to remain a meadow and not return to woodland. This is one of several long-term studies by the Department of Landscape, University of Sheffield, led by the dynamic Hitchmough. He is a scientist and sets about his experiments in a detailed and scientific way such as most of us would shy away from imitating. But he is also clued up on landscape design and the socio-economic side of the question and he is a keen plantsman. So his conclusions are well founded. He is quite outspoken, too, and gives short shrift to certain examples of narrow-minded ecological dogma.

At Harlow Carr, on a relatively cold, wet (about 250mm/10in more annual rainfall than my East Sussex garden receives), north-facing site, Hitchmough used the 'burn-off' method (described above) to clear the site, followed by sowing a native wildflower mix, which included five well-behaved native grass species. The introductions of perennials he is trying out are of hardy species mainly of European origin. This is because European herbs are better adapted to coping with the year-round green-meadow matrix that the British climate gives rise to than are American prairie perennials (see Chapter 10), which are adapted to a far harsher winter climate and to hotter summers.

Before the meadow was cut (in late autumn), each of its added contents (there were twenty species, all of European origin) was cut to the ground and the severed tops collected, each species separately. They were then oven-dried and weighed. Comparison of weights from year to year revealed which species were weakening, which were holding their own and which were actually getting larger – which is what is needed. Those that have more than held their own are *Persicaria bistorta* 'Superba' (bistort, which loves wet conditions); *Geranium psilostemon* (which semi-failed with me, but I started it off without any cultural advantages of a weed-free area around it); *Thalictrum aquilegiifolium* (which would pretty well have died off by mid-August anyway); *Trollius* 'Canary Bird' (spring-flowering globe flower; another perennial at the end of its growing season) and, most successful of all, *Euphorbia palustris.* These are listed in order of increasing success.

Conditions in different meadows, not to mention in different climates and parts of the country, will yield different results. At the Royal Horticultural Society garden at Wisley, in Surrey, where summers are significantly hotter and rainfall lower, similar experiments are being conducted with tufted grasses, suitable for prairies.

OVERLEAF Jane King's meadow at Pentridge House in Dorset is on chalk downland, which suits the yellow cowslips (*Primula veris*) that carpet the area in late May among annual hay rattle (*Rhinanthus minor*), quaking grass and ox-eye daisies just coming into flower. The calciferous soil also suits the bee orchid (*Ophrys apifera*) and the pyramidal orchid (*Anacamptis pyramidalis*), both of which have appeared during the last ten years. The meadow is managed by Mrs King's ponies (see page 106), which graze selectively, eschewing precious orchids as well as the cowslips and ox-eye daisies.

5. Meadow Management

A MEADOW CANNOT JUST BE LEFT to take care of itself. Not only will it become unbearably coarse, tussocky and unsightly, but it will soon be invaded by trees and cease to be a meadow. Its care aims at keeping the sward fine, so that sweet grasses and less competitive plants can thrive, but without impeding their growth cycle, including self-seeding. To this end, when you do cut an established meadow you want to make a job of it, first removing the coarse herbage, then getting the turf really short – almost as tight as a lawn, in fact. In both cases the cut herbage should be removed (we compost it).

Clearly, the operations are laborious and labour costs (even if they are your own, and unseen) are high, so machinery is needed to make the jobs quicker and easier. Burning is a way to clear in one sweep, but it is a risky business requiring a professional and is prohibited in some areas. The benefits are that it discourages woody seedlings from taking hold. However, the ash encourages grasses to grow larger.

Autumn and winter grazing is kinder to wildlife but has its drawbacks. It can be useful for controlling the coarse competitive grasses, but trampling of the vegetation and heavy 'poaching of the land' may occur. Grazers can be selective, choosing only the most palatable plants, putting these under pressure while others, such as ragwort (*Jacobaea vulgaris*

Meadows contain an inordinate number of grass-loving insects and spiders that satisfy the needs of ground-dwelling birds. Spiders' webs attached to long meadow grass show up best on dewy autumn mornings when the light catches the threads that compose their complex designs: the outer threads forming the framework are the strongest; those at the centre are sticky to catch the prey.

syn. *Senecio jacobaea*), which is poisonous to horses, and buttercups (*Ranunculus* spp.), are allowed to take over. Animal urine is high in nitrogen and potassium and droppings contain large levels of phosphorus and calcium, which result in 'soiled' patches avoided by the grazers where nutrient-demanding, vigorous species develop, such as creeping thistle (*Cirsium arvense*) and nettles (*Urtica dioica*).

Cutting is more controlled, less selective, does not 'soil' the ground, and doesn't trample or poach it, but feels less holistic. Grass-cutting machinery has made meadow management far simpler. Fergus Garrett used to love the physical action of using a scythe effectively, but even he had to give in to the speed and superiority of a range of mechanized tools. Some of them are sit-on, although we have none of the large areas that make this way of doing the work wholly effective, but a man really feels a man when he's up there, above the rest of us, managing a machine, and that needs to be taken into consideration.

The meadow at Pentridge House (illustrated at cowslip time on pages 102–3) is managed entirely by selective grazing – Jane King's ponies being suitably picky about what they eat.

WHEN TO CUT?

When and how often to cut – these are the questions that most worry the meadow gardener. There are no hard-and-fast rules. At Knowsley in Merseyside the meadow is cut just once a year, at the end of summer. At Harlow Carr in Yorkshire, Hitchmough's annual cut is done in late autumn and is so tight as almost to skim the area, leaving bare patches, as would have been the case had it been grazed with sheep and cattle. These open spaces give their chance to self-sowing meadow contents. At Sticky Wicket in Dorset, Pam Lewis cuts between mid-June and the end of August, mowing coarse grassy areas first, delaying the cutting of those where wildflowers are flourishing until they have set seed. She leaves some patches of grass for insects. She grazes or mows the regrowth in autumn and again in spring if it is lush.

At Dixter, the time factor is crucial because of the presence in most of our meadows of autumn- and spring-flowering bulbs, without which, to my mind, their joy would be considerably reduced. The late-summer cut is fairly urgent in areas where there is a little season of flowers from colchicums and autumn crocuses. We used to start cutting our meadows in July, with a further cut at the end of August to allow autumn bulbs to be seen and appreciated the better. But now we begin the first cutting in mid-August – not all our meadows at once, by any means; some wait till September. In fact, the cutting of long grass continues non-stop throughout the month, with the whining strimmer going into all those awkward-to-reach spots.

Timing the late-summer cut of species-rich meadows on poor soil is a question of fine-tuning. At Dixter we start cutting some areas in mid-August but delay in others until the seed pods of late-maturing common spotted orchids turn from green to brown as they ripen and split to shed their dust-fine seed.

Cutting in late summer has proved to be a good compromise, being early enough for the autumn bulbs yet late enough to have allowed most of the meadow's contents to ripen and shed their seed. To allow seeding before the first cut is very important, because your plan will be to establish colonies of plants, mostly bulbous, that, once given a start, will naturalize and thus do the gardening for you. Admittedly, it does reduce the vigour of the perennials that are still green and growing, but an even later cut encourages coarse herbage and the dominance of a coarse turf such as muscles out the smaller-growing plants. A difficult bit of fine tuning.

Turf growth in the upper moat is coarser and more vigorous than elsewhere, because it was the bottom of a pond before it was drained, so we often give it a second end-of-summer cut before the first cuts elsewhere have been completed. The second cut gives a nice low background for *Crocus nudiflorus* when that gets going, usually in mid-September. If grass growth is good, then three cuts are given: two before the crocuses flower and one after.

When we have cut down the meadows either side of our front path, visitors often ask if we have been over them with weedkiller! But the base of old grass stalks is naturally hay-coloured. Within a week, green will be appearing and in the second half of August there will be colchicums in bloom. In the orchard, which is our next big area to be tackled, the cut reveals contours of large ridges and hollows, which is the way the land was ploughed in medieval times. Although I love long grass, there is also something very satisfying in seeing everything clean again.

A second cut is needed to prevent the turf becoming too coarse and tussocky. We do it in November, after the colchicums have flowered in the front meadow and, in the orchard, before the daffodils' noses have pushed through far enough to be damaged. This tight cut late in the year makes certain the lowlier spring bulbs, such as hoop petticoat daffodils, crocuses and orchids, are displayed most visibly and enjoyably.

A third cut can be made in spring if your meadow doesn't include bulbs. This enables the climax of floral display to be in August, as John Sales, the one-time chief gardens adviser to the National Trust, does to parts of the meadow in his Gloucestershire garden. No cutting should be done after hay rattle has germinated (in February).

To sum up, then, there are no hard-and-fast rules, but timing and frequency of cutting are based on all kinds of factors. As well as aesthetics (long, bleached grasses are beautiful, until laid by wind and rain, and I have confessed elsewhere to a feeling of relief when formality is restored to the topiary lawn), timing includes such practical considerations as:

ABOVE A bush cricket on comfrey in a Hungarian meadow, one of the many invertebrates that depend on vegetation being left undisturbed until they have completed their life cycle.

OPPOSITE September in the front meadows, with Fergus Garrett tackling the long grass. First he uses the power scythe, mowing the same row twice, once in each direction. Next he rakes the hay into rows for easier collection and then he uses a two-pronged hayfork to load it into a tractor trailer for transporting to the compost heap. Lastly, with a ride-on rotary mower fitted with a collection box, shown overleaf, the grass is cut again (ideally when it is dry), to get it really short. In the new year this area is carpeted with short-stemmed bulbs, including snowdrops, followed by crocus and narcissus (see pages 156–7), so the grass is given a final cut in late autumn and it remains short through the winter.

When the latest and most important plant ripens and sheds its seed. At Dixter this depends on the late flowers such as tufted vetch, lesser knapweed and the common spotted orchid.

When the invertebrates – grasshoppers and butterflies being the most obvious – **have completed their life cycle.** At Dixter this is usually at the end of August.

When bulbs are safely below ground.

When the weather is dry, so the clippings are easier to collect. (Sometimes we have no alternative but to cut in wet weather.)

The need to control certain plants in the sward.

The height of the grass. It is better to cut long grass before it falls over, otherwise it gets tangled, looks unsightly and is more difficult to mow.

THE MECHANICS OF CUTTING

If you have very little meadow grass, a strimmer will manage the whole thing, but most of my readers will have areas large enough to warrant buying or hiring machinery. Rotary mowers are not strong enough to cope with long grass unless they are heavy-duty ones behind a tractor.

Machinery clumsily used can do a lot of irreparable damage in a very short time. If you damage a young tree trunk, there remains the scar forever more, damaging the functioning of the tree but, even worse, damaging its appearance. Fergus is well aware of this and very strict in his instructions, and also on the safe operation of all machinery. Paid labour is legally forced by Health and Safety regulations to take many precautions when mowing. Unfortunately, there is nothing to prevent the amateur from acting as foolishly as he likes. He (or she, obviously) should wear ear defenders, steel-toe-capped boots, helmet and visor and heavy-duty trousers. Using a strimmer, things fly up at you with great force. A Dutch lad, working a spell with us, had lost an eye this way. Yet the ignorant will set about the job wearing shorts and with no protection. It makes you feel queasy to see.

The final cut on the rotary mower.

Our heaviest machine is a power scythe Tracmaster (which is a trade name; there are others of different makes that are similar). It is strong (though the smallest of its family that is made) and doubles up as a rotovator. It is relatively easy to use and it has guts. It is always important to have a machine that is strong enough for the job and some to spare. You walk behind it and it cuts long grass with reciprocal blades in the same style as the old Allenscythe, but it is a great deal safer. That could run away with you and be hard to stop. This has a dead man's handle which stops the machine the moment you release it.

No second person is needed to pull the grass you have cut out of the way. It cuts the grass to 4cm/1½in, but no shorter. We mow the same row twice, straight off, once in each direction. On uneven ground, it may knock grass over without cutting it on the first cut, but this is dealt with on the return journey. The grass is raked with a wide wooden rake and is easily picked up by a two-pronged hayfork, which loads it into a tractor trailer.

It is then taken to the compost heap (the wetter it is, the better) and stacked loosely, spreading alternate layers of sulphate of ammonia and lime every 15cm/6in. It is then watered. Quick fermentation is thus promoted, and the looser the stack, the more easily is this achieved. When it has sunk and compacted, it will need turning, feeding and watering again. This will make it ready for use in the vegetable garden when aged eighteen months. I am not saying that our methods are ideal, but they work for us. The compost is not weed-free, so we do not use it on the borders.

Back to the initial job, 4cm/1½in long is not nearly short enough. We next go over the area with a ride-on rotary mower, which takes the grass down to 1cm/½in. It won't take long grass, so it has to follow the Tracmaster. It has a sweep which collects into a box, this operation being most efficient when the grass is dry. This is emptied on to a heap. You can spread hessian (burlap) or some sort of cloth over the ground first, to make collection of the heap easy. Ejection from the box on to the heap is done with a lever and involves no dismounting.

Fine mowings are awkward to collect and you might say (almost certainly will), why bother? There are several reasons. Leaving them is unsightly; you pick them up on everything, your shoes, clothes, everything and bring it into the house. They are deposited in layers shading out the grass beneath. They put nutrient back into the soil and your object at all times should be to impoverish it. However, Landlife has found that exceptions can be harmlessly made to this rule and I have mentioned this elsewhere.

You finish off near trees and in corners by hand, with a strimmer, cutting with a heavy-duty plastic cord, which cuts grass better than a blade. With an old tree that has thick bark, you can use the strimmer at very low revs and go close to the bark. Near young trees, finish off by hand, pulling the grass away from them and using secateurs or shears. This is often omitted as being too great a bore but makes a great difference. You may try to use the strimmer carefully close to the young tree but will always come a cropper sooner or later and then there's nothing to be done about it. Damaging gunnera leaves that are lying on the grass near our moat and horse pond areas, is also all too easily done. You need help from a second person to lift the leaf up. Rake up clippings with a springbok rake and pick them up between boards.

The strimmer can do everything (it makes a vile and penetrating whining noise), but cuts grass to a mush, making it hard to pick up. The Flymo can deal with steep slopes, but is less versatile than the strimmer and won't get into nooks and crannies, which the strimmer can and deal with slopes, too. On small areas, where a ride-on is out of the question, we use a smaller, walk-behind rotary mower, which collects the grass.

Low sunlight casts long shadows across the tightly mown sward in the orchard meadow. The grass is cut as close as possible in late autumn, ahead of the snouts of bulbs peeping through the turf, which will make the display illustrated on page 162.

6. Mainstay Grasses

The most competitive component of a meadow is usually the grasses. These may account for 80 per cent of the meadow's plants and there are times when little else will be in flower. Even if a meadow were 100 per cent grass, it would not, or should not, be just a blank and inert mass of green, although our aim is to create species-rich grassland. Grass has a life of its own. It has movement and it conveys the sense of a whole community of life living within and depending on it. Many grasses also have flowers – they may not be as bold or as striking as, say, some of our terrestrial orchids, but their role in the meadow scene should not be underestimated.

Perhaps the best way of discovering the living nature of grass is to lie on it, in quite early spring when the ground has dried. You need nothing but the grass underneath you to discover that it has a very strong and agreeable smell, all its own. The smells vary according to the grasses beneath you, but however that may be, the experience is very special.

As grasses make up the main population of a meadow community, it is essential that they do not become so vigorous as to oust desirable species. If you reduce the vigour of your grasses, the mixture of species will be all the richer. Some of the best meadow communities exist on chalk downland in the British Isles and on the thin soils of the European Alps. Thin soil is unfavourable for lush grass growth, allowing a greater range of plants to take their place in the community. This is illustrated by one of our areas of poorest meadow at Dixter, which we yet regard as most successful in its varied tapestry effect. Opposite the top of the long border, the grass remains short and neat until we are ready to cut it.

In most circumstances the coarse grasses are undesirable because they oust the weaker kinds, and many flowering plants likewise, but there is often a place even for them and, if they are mown frequently and hard, they may be so reduced in vigour as to be rendered harmless. The reduction in vigour has another advantage. The chief grumble against grass is that it gets laid by heavy rain and doesn't stand up again. The coarser and taller the grass, the more unsightly it looks, and the harder it is to cut. Where the grasses grow tall and flop, we are forced to cut much earlier, so the season in these meadow areas is relatively shorter.

Where excessive rankness causes management problems and has a tendency to reduce the number of species, the less aggressive are too greatly competed with by the few which over-fertile conditions enable to dominate at the expense of the rest. One good way to reduce the vigour and hence the height of grass is with the semi-parasitic hay rattle, *Rhinanthus minor*, which I have mentioned as solving the problem of grasses flopping over the path in our front meadows. We have only introduced it to our Dixter meadows of quite recent years, but it spreads quickly and its range can quickly be

Grasses are the matrix within which other members of the meadow plant community come and go through the seasons. In early summer at Ketley's, in Sussex, common sorrel, *Rumex acetosa*, makes a haze of russet and buttercups add touches of gold to a lively tapestry of green.

extended by scattering its seed, as soon as ripe, wherever you want another colony. It is an annual, so must be allowed to seed if it is to continue, but its seed is ripe in early summer. It remains viable for only a short while and requires a winter spell of cold to break seed dormancy, so the more quickly it is distributed where required to grow, the better. It has a boom-bust cycle of three to four years. It can be eliminated in one year by cutting before flowering.

I do not know all my grass species. Desirable though it might be that I should be able to identify them, I do not think this is essential (I am probably guilty of complaisance). But I know my favourites. None of these grasses should be included in North American meadows (see Chapter 10), where the introduction of European grasses, thought to have a higher nutritional value than native species, has aggravated the problems of invasion by foreign species.

AGGRESSIVE GRASSES

Perennial ryegrass (*Lolium perenne*, dear to many farmers' hearts), cock's foot (*Dactylis glomerata*) and tall fescue (*Festuca arundinacea*) are unsuitably aggressive. So is Yorkshire fog (*Holcus lanatus*), though beautiful in flower, when it is pink, as we saw it at Knowsley on Merseyside. The name, by the way, originates from the Old Norse *fogg*, meaning a 'long, lax, damp grass', but it is more commonly taken as a description of its misty

In the Derbyshire Peak District *Holcus lanatus* flourishes by drystone walls where the tractor cuts the hay less often. With soft stems and leaves densely covered with fine hairs and, in summer, pinkish panicles that fade to buff, Yorkshire fog is beautiful, but it is aggressive, especially in rich soils.

appearance from a distance. *H. mollis* – creeping soft grass – has little to recommend it and is a great spreader by vegetative means. Creeping bent, *Agrostis stolonifera*, is another to avoid. Couch or twitch, *Elymus repens* (syn. *Agropyron repens*), with its extensively creeping, wiry rhizomes, is universally detested, but is difficult to get rid of in a mixed grassland community. However, even couch will not stand close and frequent mowing, as in a lawn.

GOOD GRASSES

We now come to the non-competitive grasses that provide a matrix for wildflowers. Many are also favoured by butterflies. A basic grass mix of common bent, *Agrostis capillaris*, and fescues (*Festuca* spp., including red fescue, *F. rubra*; slender creeping red fescue, *F. r.* subsp. *litoralis*; Chewing's fescue, *F. r.* subsp. *commutata*; dwarf *F. pseudovina*; and meadow fescue, *F. pratensis*) is most desirable, but your choice of species will depend on conditions of light, moisture, fertility and pH.

The two earliest-flowering grasses are easily recognized. Sweet vernal grass, *Anthoxanthum odoratum*, is out by the end of April. Stemmy and not very tall, it is said to be largely responsible for the strongly pleasing smell of newly cut hay. Then comes meadow foxtail, *Alopecurus pratensis*, tallish with a dense, dark green spike. This prefers damp soils. When young and fresh, the top section pulls easily out of its socket, and my sister and I liked to do this and nibble the sweet tip so exposed.

Meadow foxtail, *Alopecurus pratensis*, is a tallish grass with a dense dark green spike. This prefers damp soils, forming dense tufts and spreading by rhizomes. Together with sweet vernal grass it is one of the earliest grasses in flower.

LEFT Crested dog's tail, *Cynosurus cristatus*, a grass often included in lawn mixtures, has dense, upright, wiry spikes that look braided. Its tufted habit makes it an ideal grass for a meadow. In the nineteenth century it was grown for making bonnets.
BELOW LEFT Quaking grass, *Briza media*, with parasitic field cow wheat, *Melampyrum arvense*, in a Hungarian meadow.
BELOW RIGHT Common bent, *Agrostis capillaris*, is particularly at home in poor soil. With short rhizomes and very fine growth, it has long been part of our topiary lawn, but was closely mown and never flowered until I allowed my father's former putting green to become a meadow. Its loose, open flower panicles form a fine pink haze that looks magical when laden with dew.

Crested dog's tail, *Cynosurus cristatus*, is excellent on rather poor soil and easily recognizable, with dense, upright spikes, the anthers mauve when it is flowering. It seems to be most abundant on our topiary lawn, which I allowed to become a meadow about ten years ago.

In my garden, my favourite grass is common bent, *Agrostis capillaris* (formerly *A. tenuis*). It is what is known as a poverty grass, particularly at home in poor soil. Its growth is very fine (used in fine lawn mixtures) and it is just about the latest grass in flower, easily persuading me that July is too early to be making our first cut. The flowers form a fine pink haze, particularly beautiful when laden with dew.

Anyone can recognize and love quaking grass, *Briza media*, which was plentiful, and charming as ever, in the Hungarian meadows we visited. It does not grow naturally at Dixter, being generally found on alkaline soils. Ours, derived from Wadhurst clay, hovers around neutral. In the Peak District, on poor, thin soil where there is little competition from vigorous grasses, it tolerates quite a wide pH range. Having previously associated it with thin, chalky soils, I am tempted to introduce it to my poorer soil areas.

7. Perennials

The plants you encourage to grow in your meadow, or introduce deliberately, are a matter of experiment and experience. Meadows need managing and, from time to time, they also need editing. As I have said, the plants in our meadows at Dixter vary very considerably over quite small areas. One will fade away in one area but will thrive in another. Similar localized variations occur in all meadows. In Hungary, the Peak District and North American prairies I have seen how a pocket of different soil, or a wetter or shadier patch of ground, results in sometimes remarkably diverse flora. Even an area that has been cut less frequently than an adjacent one can be surprisingly different.

Thanks to their abundance, the commonest plants, which will anyway make the greatest show, are easy to establish anywhere, even supposing they are not already present. Such are buttercups, moon daisies (or oxeye daisies) and the common grasses. Nobody need despise them.

North American meadows (as opposed to prairies) are most in evidence in the East Coast states that are typically humid in summer and where annual rainfall is 110–123cm/45–50in. Here a wonderful assortment of colourful plants moves in where cultivations have ceased – at first 'pioneers', such as ragweed and goldenrod, but over time such plants as *Aster* spp., the short-lived black-eyed Susan (*Rudbeckia hirta*), ironweed (*Vernonia* spp.) and grasses. Occasionally, meadows are self-perpetuating and will not return to forest; for instance on very wet land. This may be dominated by sedges and a few filipendulas; also *Eupatorium purpureum* and *E. maculatum* Atropurpureum Group, *Asclepias incarnata*, *Sanguisorba canadensis* and *Aster novae-angliae*. Very shallow soil on bedrock is often the worst drained, giving rise to wet meadows.

A great enemy to natural meadow, as we have seen elsewhere, is invasion by foreign species which, deprived of the natural checks in their home territory, take over in a big way when let loose in a new habitat. This is particularly so in the United States of America and has been aggravated by the introduction of European grasses, thought to have a higher nutritional value than native species.

However, British native flora (hard to define anyway, as many species that we take for granted as native were probably introduced long ago) is impoverished. After the last Ice Age, the flow of species returning northwards from Europe was prevented by the development of the English Channel, thus cutting us off from our heritage. It seems only sensible to restore some of this, especially in an urban landscape where there can be little pretence to a local or natural flora.

When a garden escape goes crazy in a country like the USA, such as the purple loosestrife, *Lythrum salicaria*, introduced from Europe, where it has so many natural checks as never to become a nuisance, there can be real trouble. Well behaved by pond and canal sides in Britain, where it competes on equal terms with many

The bee orchid, *Ophrys apifera*, attracts pollinators despite resembling a bee. It is apt to turn up in unlikely places but is seldom amenable to deliberate introduction.

other wildings of similar vigour and preferences, on finding itself in the cooler, damp regions of north America, these checks were no longer present and it took off, being more aggressive than the endemic flora. So, no kind of lythrum (even if a sterile cultivar) is any longer allowed to be marketed or planted in the USA. A case of shutting the stable door after the horse has bolted, but the authorities have to be seen to be making some sort of effort. Besides which, authorities are not always very intelligent in respect of plants and the way they work.

In a small country like Britain, trouble of this kind is less likely, though there are examples. Red valerian, *Centranthus ruber*, has colonized many areas, especially alkaline cliff faces, but it is popular. *Buddleja davidii* is generally loved for its scent and for providing nectar for bees and butterflies. That gets itself into all sorts of unlikely places (especially sites bombed out during the last world war). *Allium triquetrum*, a spring-flowering bulb with pure white flowers, has taken over most of Cornwall and other areas. It looks charming on roadsides with red campion and bluebells, but at Dixter we have, after years of trying, still failed to rid ourselves of it where it invaded from a neighbour's garden.

And what of the ubiquitous *Rhododendron ponticum* of acid soils, especially in the north of Britain? It makes good windbreak shelter when you are establishing a woodland garden and it is a beautiful landscape feature at its time of flowering, but it seeds around everywhere and takes over areas that should be pasture. Not at all easy to control. But there are so many other instances where an alien's presence seems pure gain – the snowdrop, for instance; and the snakeshead fritillary's claim to be a genuine native is questionable. I believe we need to remain pragmatic and to keep an open mind. And, as we have seen, that is James Hitchmough's view. There are, one must remember, genuine natives that are a scourge as weeds as well as exotics.

LEGUMES

It had long been recognized, even before the reasons were understood, that the presence of legumes, members of the pea family, *Papilionaceae/ Fabaceae*, enriches the soil they grow in. It was later found that the nodules on their roots enable them to fix nitrogen. In a meadow, soil enrichment is likely to be undesirable, impoverishment being far more desirable as facilitating a varied plant tapestry. Legumes are therefore often omitted from commercial meadow seed mixes. However, in old meadow turf such as we have at Dixter, the legumes are there, like them or not, and they bring so many delights that one would be loath to wish them away.

If you do introduce legumes to your meadow, make sure that they are of native origin (either seed or plants), and are not the agricultural equivalent, which will have been bred for vigour and productivity (the seed very possibly from abroad), which is exactly what you do not want.

One of the most aggressive, I suppose, because of its creeping root-stock which enables it to colonize large areas, is *Lathyrus pratensis*, called meadow vetchling. It has yellow pea flowers, agreeable enough but seldom making a great display. It is

Tuberous pea, *Lathyrus tuberosus*, makes a great display in thin turf, often on disturbed ground by roads. It has a running rootstock and can be invasive, but gets elbowed out in my dense turf. It was grown as a root vegetable in the early nineteenth century.

rather weak-stemmed and apt to lodge in wet weather. I saw it in a hayfield in the Peak District and until then had never thought of it, as the Hay Meadows Report tells us, as a poor competitor and readily lost if the land is improved. It spreads vegetatively into quite large colonies, not forming new colonies easily, as the seeds are heavy and stay where they fall. But, not being dependent on seed, it can cope with earlier cutting than plants which do. It is rich in protein and valuable in a hay crop. *L. tuberosus*, with a running, tuberous rootstock, comes a little later and is one of the showiest, with carmine red pea flowers. I saw it in Hungary and have tried it without success in my own meadow. It cannot cope with that perennial mat of herbage. We used to have the grass vetchling, *L. nissolia*. It was rather precious; I remember it particularly by the front path. The leaves are reduced to a single, grass-like blade. The flowers, sparsely borne, are intense crimson. I have seen it once since, but not for a long while.

What about clover? We have the red and the white in considerable quantity. White clover (*Trifolium repens*) is one of the most persistent 'weeds' of lawns. It remains dark green and healthy even in time of drought and manages to flower even under tight mowing regimes. It will be covered in bees which will be picked up with the mowings. Apart from not being grass, it has two disadvantages. It never makes a complete cover, always patchy, and it is apt to be slippery underfoot.

Red clover, *Trifolium pratense*, grows tall enough, up to 30cm/1ft, to cope with tall grass (unlike white clover, *T. repens*, which has a creeping habit). With unmistakable individuality and brilliant magenta colouring, which varies considerably, it is a great asset in the tapestry. Although it has little fragrance, red clover is a source of delectable honey.

None of which needs to concern us in a meadow, but it should be realized that white clover actually thrives under a tight mowing regime. If the surrounding herbage becomes long, it does less well and may be ousted under over-rich meadow conditions. Mention must be made of its delicious scent, carried on the air in warm summer weather. It flowers in high summer.

Red clover (*Trifolium pratense*) behaves altogether differently. Since we have allowed our topiary lawn to become meadow, red clover has thriven exceedingly at the expense of white, because its habit is naturally so much taller. Its magenta colouring varies considerably from plant to plant but is an invaluable colour at its best, contrasting strikingly with predominant yellow. Its high season is quite early, in May, though it goes on, to quote John Clare's 'A Sunday with Shepherds and Herdboys' (*c.* 1820):

Like a richly coloured map
Square platts of clover red and white
Scented wi' summer's warm delight.

Bird's-foot trefoil, *Lotus corniculatus*, plays an immense part in all our meadow areas, but although vigorous it is not a menace, as bought agricultural seed

Bird's-foot trefoil, *Lotus corniculatus*, grows in colonies. It has long and varied flowering times, through late spring and summer. Like all legumes it has root nodules containing bacteria. It is a larval plant of the burnet moth and the food plant of the common blue butterfly, which will generally be seen fluttering in its vicinity, remaining active on it until the end of August and delaying our cutting of the meadow.

would be liable to be. It has a running rootstock and makes colonies, dense colonies packed with blossom which is apt to come in flushes, any time from May to August and in every kind of sward. The heads of yellow pea flowers are often streaked or flushed with bronze, making a nice contrast with the common blue butterfly which feeds and breeds on it. (As I've said, we delay cutting certain areas to allow the butterflies time to complete their cycle.) I would not be without this plant. I learned in the Peak District hay meadow that it is one of the first species to succumb in meadows that are fertilized. This explains why, in my own meadows, the most dramatic and dominant colonies are on the poorest bits of ground, though it had never occurred to me to put two and two together. It has a plethora of local names; some refer to the shape of the individual flowers (lady's slipper, Dutchman's clogs), others to their colour (butter and eggs) and yet others to the long, claw-like seed pods (fingers and thumbs, lady's fingers and the particularly appropriate granny's toenails), which can be up to seven in number.

Vetches, of the genus *Vicia*, are mostly, in my area, rather boring and often invasive, especially when they find their way into flower borders. Exceptionally, *Vicia cracca*, the tufted vetch, is a most beautiful plant, semi-climbing – very handsome where it has climbed up a yew hedge – and one of the latest in flower,

Lupins make a dramatic display naturalized *en masse*, but they do bring problems of nitrogen-fixing – making the soil richer and the turf coarser – as well as shading out other meadow ingredients, resulting in a monoculture that regenerates year after year.

in July, but there are early-flowering strains of it, even in quite cool habitats. It has dense heads of near-blue flowers, a most striking colour in the predominantly yellow meadows at this time of year. You see it on motorway verges and notice that its colour varies as also its time of flowering. Meadow cranesbill, *Geranium pratense* (see page 219), is roughly the same colour, generally flowering slightly earlier. Tufted vetch has its own favourite areas. I have raised it from seed and tried to establish it in other places, with some slight success.

We saw tufted vetch in Hungary, growing in dramatic abundance in a verge, side by side with crown vetch, *Coronilla varia*. The coronilla has a running rootstock and makes a good garden plant if you can let it loose where it won't become a nuisance. I remember it in a rough, neglected bank at Wye College, in Kent, where it seemed ideal. It has whorled heads of pinky-mauve flowers which were being fed on by blue butterflies. In quantity, it was strongly fragrant on the air.

The lupin, or lupine as it is known in North America, from whence it hails, is a tallgrass prairie plant (see Chapter 10). In early summer you can admire these colourful aliens all over Northern Europe, forming colonies in sandy, acidic wasteland. Generally these are self-sown Russell lupins, a cross between tree lupins from California and *Lupinus polyphyllus* from North America. Encouraged by the deliberate sowing of sophisticated seed strains, they are a considerable tourist attraction in New Zealand's South Island, albeit at the expense of native flora. For our borders, we go for the colour-segregated seed strains, but we don't

A wonderful flowering tapestry of field scabious, ox-eye daisies and a yellow-flowered cow wheat that reduces the vigour of grasses. The field scabious, *Knautia arvensis*, an invaluable wildlife plant, grows in thin grassy places, often where it is quite dry. The flowers are packed into dense heads and the petals in the outer rows are much larger than those in the centre. The ox-eye daisies here are typical of long-established colonies where numbers are quite sparse.

grow lupins in the Dixter meadows: our turf is too dense and they would fizzle out within a couple of years.

Dyer's greenweed, *Genista tinctoria*, is a broom that we have introduced and are still learning about. Although a shrub, it will put up with being treated as a herb and being cut to the ground in autumn. It has brilliant yellow pea flowers in June–July and can spread by suckering. It is beautiful and striking and I do hope that we shall succeed with it. Easily grown from seed, we plant out the seedlings, having raised them in pots. There are roadsides where it is a familiar sight in June and I always look out for it in my favourite spots. The pungency of its colouring draws my attention even if I am thinking of something else. It was cultivated for many centuries for the yellow dye that it yields. Relics of these cultivations still exist.

SCABIOUS

The family *Dipsacaceae* includes the teazels, which need not detain us, but the pincushion shape of scabious flowers (really an inflorescence of numerous flowers) is popular, even in comparatively insignificant, wild species. All are happiest on light, well-drained soils and quickly disappear on heavy clay. The autumn-flowering devil's-bit scabious, *Succisa pratensis*, with roundish violet-blue heads and reddish anthers, is a notably adaptable species of any sort of grassland, in sun or partial shade. Field scabious, *Knautia arvensis*, with flowers in shades from pale lilac to purple, is a good butterfly nectar plant and attractive to burnet moths. The smaller *Scabiosa columbaria* is pale-flowered and confined to calcareous grassland.

COMPOSITES

The huge family of *Asteraceae*, long known as *Compositae*, gives us many of the showiest and most numerous meadow ingredients. This is the family of the daisies, most familiar of all being the common daisy, which is never happier (though generally unwelcome) than under lawn treatment. Its rosette-forming foliage sits tight against the ground, but if the surrounding herbage grows long, the daisy is in trouble and becomes smothered under meadow conditions. A daisy lawn is delightful, especially in spring, always provided you are not lawn-proud. It is pink in the mornings and on dull days, the daisies being closed and having pink undersides to their rays. When the sun opens them wide, they become white and make a terrific show. They have not had a good time of it at Dixter, in recent years, since I have allowed several areas of lawn to become meadow. And we do, from time to time, treat the remaining lawns with selective herbicide, which sets the daisies back.

The other principal daisy is *Leucanthemum vulgare*, once *Chrysanthemum leucanthemum*, called ox-eye or moon daisy. Opening in late May, it is at its best through June and particularly successful on disturbed soil, needing open ground for colonization. The plant has a strong smell, when bruised, likened to dog's turds. It is one of the first meadow flowers to colonize unsprayed grassland and has made a return to waysides after having been driven out of most agricultural grassland. These daisies seem to be relatively short-lived perennials and their numbers and locality at Dixter have fluctuated amazingly over the years. From being abundant in one meadow area for a while, they will almost disappear from there within a couple of years and it is hard to determine why. They also grow with us in the cracks of paving and in drystone walls, but here again their behaviour is erratic.

Common yarrow, *Achillea millefolium*, is another highly successful lawn weed, with a creeping rootstock and always remaining green in time of drought. When mown regularly, its feathery leaves make a fine if wiry sward. It is equally at home in long meadow grass, the flat flower heads (corymbs) being most commonly a rather dingy white. But, even in the wild, there are considerable colour variations, some of them in pleasing shades of pink, though we have never made a point of introducing any of these to Dixter. We certainly ought to.

Carlina vulgaris, the carline thistle, lover of alkaline conditions, I associate with chalk downland and also with the dunes of Braunton Burrows, in Devon. It is biennial and makes a fine garden plant if you can be bothered to raise it deliberately. We saw its dry, previous year's remains in Hungary, growing in recently cleared land.

I don't like the creeping common thistle, *Cirsium arvense*, one little bit. Its mingy mauve flowers are entirely lacking in glamour and the plant is prickly and unpleasant. So, even in our meadow areas, we treat that, when in vigorous young leaf, with selective weedkiller and we have very little of it. The seeds blow in from

Common yarrow, *Achillea millefolium*, has bright green foliage even during drought. It has a creeping habit and is the bane of perfect lawn nutters. The flat flower heads are typically white and sometimes flower into November, providing a late-season source of nectar for a range of insects.

neighbouring fields. Often they lodge on the face of a yew hedge and a fresh colony will start at the base of the hedge, spreading into meadow areas alongside. We have to remain vigilant.

The other wild thistle commonly entering the garden (it is abundant in recently coppiced woodland nearby) is the biennial spear thistle, *Cirsium vulgare*, which is a handsome brute, with deeply cut leaves. In their first year, these form a rosette, close against the ground. In their second they produce handsome thistle heads, 1m/3¼ft or more tall. I have a soft spot for this plant and am prepared to leave a few, but always hope to remember to cut and remove them before they are actually seeding. The only other thistle I would welcome in our meadows is *C. dissectum* (meadow thistle), but it is rare and largely confined to old hay meadows and fens.

A mauve-flowered composite of considerable importance is the lesser knapweed, *Centaurea nigra* (1m/3¼ft). It has a stiff, upright habit with very dark, tough stems. One of the latest meadow perennials in flower, it has small, rayless, purple, thistle-like flower heads, but not prickly. For many years, we had a few plants, introduced by my mother, of the greater knapweed, *C. scabiosa*, which is showier, the flower heads having an outer rim of rays. It is abundant on limy soils, like the nearby chalk downs, but was never inclined to spread with us (we are neutral to acid) and eventually disappeared. Definitely worth reintroducing.

ABOVE The common dandelion, *Taraxacum*, is the bane of gardeners because of its deep taproots, which sprout again if damaged. In meadows and especially along roadsides it comes into its own, making brilliant clumps of intense yellow early in the season, often at their brightest in April, though the flowers close in wet weather and at night.
OPPOSITE The autumn hawkbit, *Leontodon autumnalis*, flourishes in poor soil. It gets into its stride in late June but has a long season, the flowers opening to the morning sun and always facing it. The haze of gauzy pink is made by the common bent, *Agrostis capillaris*, also happiest in poor soil.

In Hungary we saw *C. sadleriana* (designated an endangered species), which is bold, tall (1.2m/4ft) and larger than our greater knapweed, its leaves deeply cut.

Now for some of the yellow composites. *En masse* they make some of the biggest unorganized displays. The rich yellow of dandelions creates wonderful splashes of colour. These are most commonly *Taraxacum officinale* group (but taraxacums are a minefield for the botanists). Dandelions close at night. When open by day, their colouring is almost exactly the same as that of the commonest hoop petticoat daffodil, *Narcissus bulbocodium*, which rather negates the latter's effectiveness in areas where both are growing. Dandelion clocks are beloved of all children; also of goldfinches, which generally eat them before they are ripe. In 'I stood tip-toe upon a little hill', Keats imagined:

The soft rustle of a maiden's gown
Fanning away the dandelion's down.

The seeding of dandelions into flower borders is a frequent nuisance; their taproots develop quickly. If broken, not lifted entire, they grow again. Spot treatment with herbicide is effective and is also frequently practised in lawns. It was a great relief, when we allowed our topiary lawn to become meadow, to be able to contemplate the dandelions with pleasure instead of resentment.

The dandelions are followed, through the summer, with a whole host of yellow hawkweeds (*Hieracium*), hawksbeards (*Crepis*), which my flora (Clive Stace: *New Flora of the British Isles*) informs me are doubtfully distinct from *Hieracium*, and hawkbits – *Leontodon*. Thin, poor soil suits the rough hawkbit, *L. hispidus*, which cannot cope with the competition on richer soils, whereas the more adaptable common cat's ear, *Hypochaeris radicata* – another of those yellow, dandelion-like daisies – I have seen in abundance, coping with the competition on richer soils. The only *Leontodon* which I can with any certainty distinguish, at least in my own garden, from the rest is *L. autumnalis*, the autumn hawkbit. Its colouring is a slightly less brash yellow and it flourishes on some of our poorest bits of turf. Starting to flower in June, it runs on and often makes a return appearance after the long grass has been cut, at which time it is a favourite with late hatches of the common blue butterfly. A nice essay in colour contrast.

These yellow hawk-things tend to have leaves in basal rosettes, so they are great denizens of lawns. As soon as you allow the lawn to become meadow, there they

When suited (I wish I knew how to suit it) the orange hawkweed, *Pilosella aurantiaca*, makes incredibly individual colonies of its burnt orange daisy flowers. It seems at its happiest in the cooler north and west of the British Isles and I have seen it making a show on a roadside in Vermont. I am still trying to establish it in my meadows.

are, making a terrific show, concentrated on the period after 9.00 (BST) in the morning and closing around 2.00 p.m. But timing does vary. They face the sun, so, to admire them, you need to have your back to the sun. These flowers are tremendously cheerful and are no trouble to introduce (if necessary) and establish.

Orange hawkweed is a beautiful burnt orange, of a colouring quite distinct from that of other meadow flowers. It was *Hieracium aurantiacum* but has been put into another genus and is now *Pilosella aurantiaca.* You see sheets of it in Scotland, particularly. It was originally an introduced species from continental Europe. I long to establish it at Dixter and we are still trying. As I write, we have a couple of trayfuls of seedlings waiting to be transferred to meadow, but so far we have had little success.

Common ragwort, *Jacobaea vulgare,* is a notifiable weed. Dried and in hay, it is poison to cattle and horses, its local names of mare's fart and stinking Willie saying it all. Although a far from subtle shade of yellow, it makes a great display in July, rather later than most meadow flowers, and it has a pretty cut leaf. It is, moreover, only a biennial. We get a few plants which I allow to flower, pulling them out or breaking them off before they seed.

Golden rods, *Solidago*, mostly hailing from North America, I have so far failed to establish in dense meadow turf. We do have the native, late-flowering *S. virgaurea* in shady places, where the turf is thin. That needed no introduction. I cannot say that it makes any sort of a show.

Liking similar conditions is the leopard's bane, *Doronicum pardalianches* (1m/3¼ft), which is abundant in the north of England and south Scotland. It grows with me but did not flourish. Now, in just one area, it has started to do so and is making quite a dense colony. Taller and later-flowering than most of the doronicums cultivated in borders, the yellow daisy flowers are also rather smaller.

Tansy, *Tanacetum vulgare*, typical of the vegetation of a wet meadow, has ferny foliage and golden button-flowers. We saw it in Hungary, growing with comfrey, *Symphytum officinale* (popular with bees), whose colouring and flower size varies greatly, the foxtail sedge, *Carex vulpina*, and a 1.2m/4ft-tall fescue, *Festuca arundinacea.*

One of the tallest yellow daisies, which we grow in rough grass to make a splendid meadow effect in July, is *Inula magnifica* (2m/7ft), from the Caucasus mountains. The 15cm/6in daisies have long rays that quiver in the breeze. The grass around their widely spaced clumps is cut shortly before they flower, but is full of pheasant's eye narcissus in spring. We boost the inulas by mulching them in winter.

Inula magnifica is not a natural meadow ingredient but clumps of it planted into an area of rough grass make a striking incident in their July season, growing more than 2m/7ft tall, although some stems drop to near horizontal under their own weight. We cut the grass (mostly couch) beneath them just before this happens. On pages 76–77 the inula's emerging foliage can be seen among cow parsley, buttercups and poet's narcissus.

BUTTERCUPS AND NEAR RELATIONS

Perhaps an even brighter uninvited yellow display than that of the daisy-like hawks-this-and-that is made in our meadow areas by the buttercups, their advantage being that they are not greatly affected by the weather or the time of day. One of the most enlivening aspects in buttercups is the lacquering of their petals, which puts a gleaming shine on them. Children love to walk though a field of buttercups – I did, anyway – and to see the coat of yellow pollen that they lay on one's shoes.

We have four species of buttercup growing wild at Dixter; five, if you include celandines, which share the same petal lacquering. The celandines have already reached a flowering peak by late March, but they are extremely light-and-warmth-sensitive, remaining shut if conditions are not to their liking. They are very much at home on our stiff clay soil and enjoy the wettest areas. The oldest form of double celandine is just as happy in grass and is a long stayer. The flowers are very neat double rosettes, packed with petals, and they have a green centre which suits them admirably. If your garden is plagued with pheasants, usually survivors from a local shoot, they will eat every celandine flower, and also the snakeshead fritillaries.

First of the true buttercups to flower is goldilocks, *Ranunculus auricomus*. Fergus and I both have a special affection for it, based on its excellent qualities. It is a

The old-fashioned tight double-flowered celandine makes glittering yellow rosettes with a green eye. It is a great survivor and often to be found in dense grassland among coarser elements. Like other celandines, it loves heavy, water-retentive soil.

Caltha palustris var. *palustris*, often misnamed *C. polypetala*, is a large-flowered kingcup, more inclined to grow actually in shallow water than merely in boggy places. Furthermore, while the typical kingcup, *C. palustris* itself, is clump-forming, *C. p.* var. *palustris* has longer, jointed stems and is apt to lean on its elbows, the joints (nodes). From these it makes new roots and spreads further afield into colonies. Showy early in its season, starting in March at Dixter, it is inclined to hide its blossom under its own foliage later on.

deciduous species, appearing and starting to flower in March, at its peak in April, but vanishing completely in summer. It loves heavy, damp soil and is especially pretty where lady's smock, alias cuckooflower – *Cardamine pratensis* – with heads of mauve, cruciferous flowers, is growing among or near to it, that also loving wet soil. Goldilocks has small, rather cupped buttercup flowers of an intense shade of yellow. It grows in colonies and a patch can cover quite an area of ground. There are different strains of this around and one of its strange characteristics is that a petal or several petals (out of five) are often missing. Sometimes there are none. Luckily our strain is entire and makes the maximum display. There is a wonderful freshness about goldilocks.

Ranunculus bulbosus (30cm/1ft) is the next species to get going, often at the same time as red clover, which makes a striking contrast. It gets its name from the bulbous, swollen stem base. Another easy point of recognition is the way its green sepals are reflexed, right back to the stem. It is a showy and widespread buttercup, but only half the height of *R. acris*, the meadow buttercup, which follows next and is predominant in our meadows. Its flowers are on the small side but showily borne on a widely branching plant. The cultivar called 'Stevenii' is half as high again, with larger, semi-double flowers. I daresay it would establish in turf but my one effort here failed.

Finally, at Dixter, comes the creeping buttercup, *Ranunculus repens*, which makes abundant stolons and roots at every node. This makes it a pestilential weed in flower borders and it is not easily eradicated in lawns. If one could establish its double form in a meadow, that would look gorgeous, but there always seems to be the ordinary type-plant ready-established there. It loves wet places.

Nearly related and again with lacquered yellow flowers is the kingcup or marsh marigold, *Caltha palustris*. That nearly always grows by a pond side but is equally successful in any boggy spot, if I could just find one where the caltha would not be in the way when we were ready to make a summer cut. The leaves are orbicular, the flowers larger than any true buttercup, and the tightly doubled form makes a terrific spring display.

Caltha palustris var. *palustris*, long given specific status as *C. polypetala*, comes from the Mediterranean but is an extremely easy garden plant and I have a colony of that almost too well established in a wettish hollow. Its large orbicular leaves tend to obscure all else but it is impressive on first flowering, in April, with extra-large flowers. Its flowering stems are jointed, eventually reaching a horizontal position and rooting at the joint, so large colonies are made.

ABOVE Our native windflower, *Anemone nemorosa*, is typically white, though changing to blush pink as it fades. 'Lismore Blue' is one of the late flowerers, at its best towards the end of April.
OPPOSITE The early spring-flowering anemones, which open to the sun, are a great joy. They thrive in dappled shade, even in quite dense turf.

Trollius europaeus, the globeflower, is commonest as a wilding in the north of England and Scotland. I planted it in a wet place among dogwoods, *Cornus alba*, but is has merely lingered, not enjoyed itself. Probably there is too much competition from dense grass. In the wild, Richard Mabey tells me, it comes no farther south than the Derbyshire Dales and the uplands near Oswestry in Shropshire.

Dense grass is certainly the problem with another member of the ranunculus family that I have signally failed with, despite many attempts in many different sites, the winter aconite, *Eranthis hiemalis*. It should make carpets of blossom, in February, especially in the neighbourhood of trees, as it does in the light, sandy soil in Beth Chatto's wood garden. On one occasion I direct-sowed masses of fresh seed. It germinated the next spring, just a pair of shiny seed leaves. The next year it produced a true leaf but then petered out. You have to take such disappointments in your stride.

The wild, woodland windflower, *Anemone nemorosa*, does extremely well in Beth's grassy wood, as it does for me in any shady piece of meadow (the fact of its being locally abundant gives a clue) and it is not averse to an open situation either. The flowers are white when young but in old age fade to pink. They associate especially well on the grassy bank above the lower moat, fairly heavily shaded by oaks on the other side of the moat in summer, but all quite light in spring, with primroses, goldilocks buttercups and a patch or two of early orchids (*Orchis mascula*). But they are in other meadow areas in various parts of the garden and I start looking for their first blooms in early March. If fully exposed to sun, they will later be shaded by long grass. Of the common names, wooden enemies is the funniest, although the least appropriate.

The blue-flowered form of *Anemone nemorosa* that I grow in the turf near to the horse pond is called 'Lismore Blue' and comes from the Devonshires' property in south Ireland. These anemones throw up many blue-tinted forms both in Ireland (for example, *A. n.* 'Robinsoniana') and in Cornwall. I particularly like 'Lismore Blue' for the sprightly shape of its open flower. It is late-flowering and is sometimes still in bloom in May. When groups become congested with their own rhizomes they cease to flower freely, but that also applies to our normal types. Nothing for it, if you can be bothered, but to split and replant, which can be done any time in spring, while the turf is still short.

I have not succeeded with *Anemone blanda* in its rich blue form, though Mary Keen is very successful with it in her Gloucestershire garden. 'White Splendour' is so vigorous that I should have a go with that. *A. apennina* ranges in colour from white, through pale pinkish mauve, to light 'blue'. I have never seen it as intense a blue as is possible with *A. blanda*. *A. apennina* has numerous petals and is a charming plant, which my mother started in her Botticelli garden, former site of the upper moat. It has thriven ever since, seeding into the lowest part and making clumps, though never large colonies. *A. canadensis*, the white meadow anemone of North America, flowers in late spring or early summer; *A. cylindrica*, the candle anemone or thimbleweed, has greenish white flowers.

Pasqueflower, *Pulsatilla vulgaris*, with greyish, pinnate foliage and bells of purple petals surrounding a tuft of yellow stamens, is a nationally scarce plant. If your garden includes thin turf overlying chalk or limestone, this is the spot in which to naturalize it. I have tried but failed. It makes a tussocky plant and perhaps dislikes our mowing techniques. In America the same common name is given to *P. patens*, where it is one of the first and most spectacular prairie flowers of spring in the northern tallgrass region.

Maybe our mowing regime is the reason for *Aquilegia vulgaris* failing, to date. We should try again, as columbine sows itself abundantly in the borders. Perhaps if I found it thinner turf, without so much grass competition, it would be happy. After all it is often seen in light woodland.

ORCHIDS

Conservationists might prefer me to keep quiet about my past in respect of wild orchids, but I prefer to be open about it. Their strange beauty has wide appeal. My mother and I adored them, no matter how small or insignificant. To love is to wish to own (until you have matured sufficiently to know better). We wanted those orchids in our own garden, predominantly in the meadow areas.

Small-flowered terrestrial orchids such as are found in meadows can make a great display when occurring in quantity, but many of their natural habitats have been destroyed. Gardens and roadsides can become sanctuaries.

RIGHT, TOP The white form of spotted orchid, *Dactylorhiza fuchsii*, has a huge distribution with many habitats. With us in Sussex it flowers in June and its seed heads should be left until August to ripen and disperse their light seeds.

RIGHT, BOTTOM The twayblade orchid, *Listera ovata*, has modestly green flowers but proclaims itself with its opposite pair of broad, basal leaves from which it gets its name – the 'twa blades' in Scots.

FAR RIGHT The early-purple orchid, *Orchis mascula*, is a showy pinkish purple and smells disagreeably of tom cats. It usually flowers in April.

Now this I can tell you, from experience. It is no use trying to naturalize orchids that are not native to the area where you live. We have tried and failed over and over again. If you knew better from the start, you wouldn't waste time and precious material. For example, the fragrant (*Gymnadenia conopsea*) and pyramidal orchids (*Anacamptis pyramidalis*) are abundant on limy soils on the North and South Downs, not 48km/30 miles from where I live. But those chalk soils are utterly different from my Wadhurst clay. In the USA, supposing you could find a legal source, there is no point in trying the western prairie fringed orchid, *Platanthera praeclara*, unless you live in the western tallgrass region west of the Mississippi.

A probable reason for their not transferring is that orchids need a microrhizal association with some fungus, which will abound in the locality where they are found, but not outside it. Both the fungus and the plant need the presence each of the other, in order to survive.

For similar reasons, the bee orchid, *Ophrys apifera*, is not for us, although it can be frequent on short calcareous grasslands and does appear in damp clayey meadows. It is worth going out of one's way to see it in flower – like pink-winged bumblebees.

The lizard orchid, *Himantoglossum hircinum*, is a tall (up to 1m/3¼ft), bizarre-looking species which is rare in the British Isles but common on the north coast of France. Its light seeds have blown across the English Channel and established in sand dunes on the Kent and Sussex coasts. Let them get on with it.

With our local orchid species, we have done fine at Dixter. My mother and I would sally forth with fern trowels and a large trug basket between us, in quest of our booty. We knew what to expect of their root system, which seldom goes deep, and we dug carefully so as to include not just roots but also the enveloping wadge of soil. If we could find the plants before they were flowering, so much the better, but this was not essential. There were at that time, in the 1920s and 1930s, no laws against this practice. In retrospect I have, apart from the inevitable and unnecessary wastage noted above, no regrets over what we did and achieved.

First in flower is the showy early-purple orchid, *Orchis mascula*, mentioned by Shakespeare in Ophelia's death garland:

> Long purples,
> That liberal shepherds give a grosser name,
> But our cold maids do dead men's fingers call them.

Orchis means testicle – hence dog's-stones, the 'grosser name' which refers to the pair of root tubers. The early-purple's basal rosettes of leaves, often heavily and handsomely purple-spotted, are already visible in January in my part of the country, though it flowers,

The green-winged orchid, *Anacamptis* (syn. *Orchis*) *morio*, has no green that I have ever discerned in its flowers, which are a cool shade of purple. It can, when suited, be abundant in May, given undisturbed meadowland on a poorish soil. Stock acquired should be of local provenance where the fungal micorrhiza with which the roots associate are present.

conspicuously, in April, with spikes of rather reddish-purple flowers. This holds its own in various parts of my garden, sometimes increasing, sometimes dying out. We have it in both open and somewhat shaded situations, as you find it in the wild, occurring mostly in non-acidic soils. One of our most successful and lasting locations is my mother's Botticelli garden, site of the drained upper moat. It mixes with Spanish bluebell. The smell of the flowers is pungent and disagreeable and has been likened to tom cats.

This is followed by the green-winged orchid, *Anacamptis* (syn. *Orchis*) *morio*, which is our greatest success. It is of slighter build than the early-purple and its colouring is a cooler shade of purple, not flushed with pink (and not showing any green either). This is a species entirely of undisturbed, short grassland, 'formerly frequent but now greatly reduced and local', as my flora informs me.

The obvious reason is that nearly all permanent pasture disappeared during and soon after the Second World War.

The sites from which we dug up our quarry have long since vanished (some, by the local railway line which became defunct, were soon covered by scrub). Do I need to point out that ours was, in the event, an act of conservation? In the days when we cut our meadow area in June, the green-wingeds were unable to ripen their seed. When I took over, and we delayed till August, so as to give these and the even later spotteds time to complete their cycle, the situation was transformed and we can now count (if we have the time), both *Anacamptis morio* and *Dactylorhiza fuchsii* in thousands, spread over many areas into which they self-sow, for their seed is so light as to blow in from a considerable distance. The topiary lawn, which we allowed to become a meadow about eight years ago, has already acquired a healthy population of both orchids, without our lifting a finger to promote them. The process continues. All very encouraging, not to say exciting.

The spotted orchid, *Dactylorhiza fuchsii*, has a huge distribution with many habitats. The spike shown here is typical of the wild strain growing abundantly in my meadows, where it makes a stunning display in June. But there is considerable variation in colour and vigour. This orchid will make clumps. If a particular plant or clump takes your fancy, you can quickly work up stock of it in a cultivated flower border. Its seed heads should be left until August to ripen and disperse their light seeds.

Anacamptis morio flowers over a long season, peaking in May. It coincides with the twayblade orchid, *Listera ovata*, which commonly grows on calcareous sites. However, we have a number of colonies and the plants are more numerous than you would at first suspect. Initially, it is quite hard to spot them. Then you see one, then another and before long you realize that there are a great many. The plant's most conspicuous feature is its opposite pair of longitudinally ribbed leaves. Between these the flowering stem rises, but as the flowers are small and green, they do not draw attention to themselves. Still, it is nice to have them there, growing away happily. I must have acquired the twayblade from one of our local woods, where there were small colonies, but it is perfectly happy in an open situation and will sometimes build into clumps.

The frog orchid, *Coeloglossum viride* (syn. *Dactylorhiza viridis*), grows in short grassland, mainly on chalk and limestone, and we failed with it at Dixter. Less like a frog and more like a diminutive twayblade, it would be even harder to spot. My mother spotted it easily on the downs near Winchester, but I had to confess (I was only a child!) that I hadn't 'frog orchid eyes'.

Making greatest impact as a highly visible meadow feature is the common spotted orchid, *Dactylorhiza fuchsii*, which we once knew as *Orchis maculata*. Our local strain of this has heavily spotted leaves with an emphasis on transverse spotting. It is extraordinarily variable both in colouring and vigour. Mauve is the general colour, but some of our largest are near to purple and very showy. If transferred to a border, they soon build up into clumps, which can be divided. When I took it to a Royal Horticultural Society show, the relevant committee said it was too large to be *D. fuchsii* and it was sent to Kew (Royal Botanic Gardens), which pronounced it to be the hybrid *D.* × *grandis*. However, I would not trust that identification. In the wild, around us, *D. fuchsii* occurs mainly in coppiced woodland, but in my garden it is prolific in open or partially shaded meadow areas, seeding itself from one to another.

We have a few southern marsh orchids (*Dactylorhiza praetermissa*), which hang on but do not increase, though one has recently installed itself in the topiary lawn area. We also, though without lending a hand in any way, have an August-flowering helleborine, but that is hardly in meadow, being in considerable shade.

I should not normally expect the greater butterfly orchid, *Platanthera chlorantha*, to last for many seasons here at Dixter, but those from a local community, under threat from a conifer plantation, survived for many years until badgers destroyed the site by using it as a latrine, after which hogweed moved in.

There was a spell when we had, unintroduced, the autumn lady's tresses, *Spiranthes spiralis*. It was quite frequent for a while, both on our lawn tennis court (now a car park) and by the side of the front path, but it vanished as it came. The flowers, which are white, grow in a near-perfect spiral around the short stem. Where the conditions are right – short, damp turf and, probably, more chalky soil than ours – it can grow as thickly as grass.

It is always a pleasure, even a thrill, to find the deciduous adder's tongue fern, *Ophioglossum vulgatum*, barely standing clear of the grass on its return in late spring. It is most widely established on pieces of old meadow, of which, alas, there is an ever-decreasing amount.

FERNS

Meadows are not generally associated as the natural habitat of ferns, but in a number of our old meadow areas we have the adder's tongue fern, *Ophioglossum vulgatum*. Once common, it is now severely localized because of the elimination of formerly unploughed meadow. It is deciduous, reappearing each spring in April; colonies of a single, glossy, upright blade, pointed at the tip and bright green. It is only 5cm/2in high when first spotted but grows to perhaps 15cm/6in along with the grass around it. At the centre of a fertile unit is the 'tongue', which will produce ripe spores in due course. I don't know whether we have more of this little fern now than in the past, or whether we (especially a highly observant young friend, Pip Morrison) have developed adder's-tongue-eyes, so to speak, but we do seem to have a lot more of it in various scattered locations. Maybe our turf management, cutting rather late, has helped.

The adder's tongue's relative, *Botrychium lunaria* (moonwort), is equally scarce in dry upland pastures in north and west Britain, but there is nothing I can do in Dixter's heavy soil to help it along. Its blade is fringed with half-moons and was once believed to be capable of unshoeing horses and opening locks. You can see a depiction of it on a canvas-work panel at Hardwick Hall in Derbyshire.

MOSSES AND MOSS-LIKE PLANTS

I know very little about mosses but I do appreciate their beauty. They are at their liveliest in spring, when warmer weather combines with plenty of residual winter moisture. They often flourish where grasses are having a hard time of it, that is in poor soil where drainage is inhibiting other plant life. Also beneath deciduous trees, where there is plenty of light up to the end of April, the mosses breed and proliferate. After that, as the shade increases and the ground dries out, they go to rest, turning yellowish. But they have been a joy in their season.

There is still a widespread conviction among conventional gardeners that moss is a no. 1 enemy. This depends on your attitude. If you are interested in a grass lawn, rather than a meadow, moss in quantity is generally a symptom of conditions being wrong. Rather than tackle the cause of the symptoms, so that grass rather than moss will naturally prefer the lawn turf, the instinct is to tackle the symptom itself, with moss-killing chemicals. The moss is consequently blackened and looks hideous for quite a long while. Fertilizer is included with the moss killer so as to encourage the grass, and order may be restored in time for the summer, but moss is sure to return the following winter and spring.

Under meadow conditions we can leave such anxieties behind, taking a benevolent view of mosses. If they are a symptom of exceptionally badly drained ground, we may, for the sake of other meadow plants, take steps to improve the drainage. The mosses will get along somehow. But if we decide that a wet meadow has its own charms and possibilities, we may accept the situation and plant accordingly.

Pearlwort, *Sagina procumbens* (*Caryophyllaceae*), has practically no height at all and is often mistaken for moss. It makes a greenish mat of no distinction and its flowers are green and generally unnoticed. In our context, it is most obvious as a lawn weed, which benefits from mowing. Once mowing ceases, so does any consciousness of pearlwort.

THE PINKS

Caryophyllaceae, to which the moss-like pearlwort belongs, is notably the family of the pinks, *Dianthus*, and the species that I should most like to establish is *D. carthusianorum*, which I have greatly admired in European meadows. At some 40cm/16in, on stiff stems, *D. carthusianorum* produces domed corymbs of small magenta flowers in June–July. The colour is just what I need at that time of the

Ragged Robin, *Lychnis flos-cuculi*, is especially prevalent in wet meadows. Its light clear pink colouring proclaims it from afar. To see it in quantity, as I have in Germany, is a great experience. So far, attempts at introducing it to our own meadows have not been successful. There is an albino form which I cannot wholeheartedly recommend. It is the fresh pink among greens that really matters to me.

year as a contrast to so much green and yellow. But although I have twice raised it from seed and grown good plants on to put out in our meadows, it has never lasted. I feel that someone somewhere in England must have been successful, but I need to know how and where.

The smaller-growing and less intensely coloured *Dianthus giganteiformis* subsp. *pontederae* looks pretty good when massed. *D. deltoides* (maiden pink) and *D. armeria* (Deptford pink) can be found in a few scattered areas of dry grassland.

We have plenty of red campion, *Silene dioica* (80cm/32in), in our woods, where it flourishes especially (along with bluebells) after coppicing, but I have never introduced it to the garden, neither has it introduced itself. Beneath the margin of trees would be its obvious locale. Apparently it hybridizes with its close relative white campion (*S. latifolia*) to produce intermediate-coloured flowers. White campion itself is an annual or biennial herb.

Even more desirable than red campion, because its flowers are such a bright, clean shade of pink, would be ragged Robin, *Lychnis flos-cuculi* (50cm/20in) – so called because of its much divided, thread-like petals. This likes marshy places and is perfectly happy in full sun. We have what would seem to be suitable places and have grown it from seed, transplanting, but so far have had no lasting success. What would seem to be the most suitable area, near to the horse pond, is dominated by rushes (*Juncus* spp.), which are so competitive as to oust all else.

Soapwort, *Saponaria officinalis* (80cm/32in), we have (as everyone else does) in its loosely double-flowered form, 'Rosea Plena'. It will grow almost anywhere but is a redoubtable spreader and takeover plant, so I have eliminated most of that, but there are waste places where, with rather rank herbage, it is suitable. Its flowers, at midsummer, are light pinky-mauve.

SORRELS AND DOCKS

Now we come to the family of the sorrels and docks, *Polygonaceae*. Common sorrel, *Rumex acetosa* (1m/3¼ft), is indeed one of our commonest natural meadow ingredients, making a delightful red haze in its summer season. The sunnier its position, the warmer its colour. It does seed into our borders and has a deepish root but is not too difficult to extract.

LEFT The common dog-violet, *Viola riviniana* (up to 20cm/8in) is a colonizer of lawns where herbicides are not used. The flowers are unscented but their nectar is attractive to hoverflies and bumble bees.
BELOW Lady's smock, as I know it, alias cuckooflower, *Cardamine pratensis*, flowers in ditches and damp meadows in April. The best colonies are quite a strong shade of mauve. It is a crucifer and a food plant of the much-loved orange tip butterfly, which is on the wing in April and May.

The nearest natural habitat to Dixter that I know of for sheep's sorrel, *Rumex acetosella* (20cm/8in), is the shingle, reclaimed from the receding sea at Dungeness, 40km/25 miles away. It likes bare places and, in bloom, makes a really bright haze with its midsummer blossom. As a weed it can be one of the most ferocious spreaders, its preference being for poor, acid soil. I have had my troubles with it when introduced in peat used for my John Innes potting composts. From the container in which it develops, at first unnoticed, it spreads quickly into the ground around.

The broad-leaved dock, *Rumex obtusifolius* (1m/3¼ft), is a pain in the neck for meadow enthusiasts, being outrageously coarse. It occurs most frequently where the ground has been disturbed, or wherever the soil is rich enough to encourage coarse plants. Within the dense turf of an established meadow, it is seldom a problem. On no account should these docks be allowed to seed. Spot treatment with herbicide on strongly growing foliage is effective. Whereas this hefty dock thrives in pockets of rich soil, the narrow-leaved curled dock, *R. crispus*, indicates impoverished soil and is an attractive member of the meadow community. It can reach 90cm/3ft, and a single plant can produce 30,000 seeds in one season.

VIOLETS

Within *Violaceae*, I have two violets. *Viola odorata*, the sweet-scented violet, is established at the foot of an ash tree. I should introduce it to other such semi-shaded spots. There is quite a range of colour forms, my favourite being the white. They flower in the main from late March to early April. *V. riviniana*, the common dog-violet, is a few weeks later, and is in some shade of mauve. The only bit of meadow it grows in has very poor soil but is sunny. For dry banks and driveways there is *Viola pedata*, the bird foot violet.

CRUCIFERS

There are few meadow plants among the crucifers – *Brassicaceae* (*Cruciferae*). The important one at Dixter is cuckooflower, which I have always known as lady's smock, *Cardamine pratensis* (60cm/2ft). Its heads of mauve flowers (a good shade of mauve, in our local strain) are at their peak through April, often March, and it loves to grow in heavy, wet soil. Many clones are sterile, but reproduce from leaves which drop off and root. It contrasts ideally with the rich yellow of *Ranunculus auricomus*, our earliest buttercup and another moisture lover.

I have also introduced the slightly later-flowering *Cardamine raphanifolia* (30cm/1ft), which flourishes in a wet ditch and looks singularly like watercress, in its leaves. The flowers are a bit larger and distinctly on the pink side of mauve, but it is apt to be shy-flowering, though it has its on-years. I suspect there may be freer-flowering strains than mine.

OPPOSITE The spring-flowering sweet-scented violet, *Viola odorata*, has many colour variants, the most typical being violet. My favourite is the pure white albino, *V. odorata* 'Alba', which seeds into paving cracks and, to quote Fergus Garrett, 'puts a smile on your face'. Along woodland edges is its commonest habitat, straying into semi-shaded meadow areas.

PRIMROSES AND COWSLIPS

Top of the pops in the *Primulaceae* is the primrose itself, *Primula vulgaris* (15cm/6in), which abounds in the woods around us. In the garden it is happiest in thin turf within some shade from trees, especially on the banks overlooking our lower moat, where it makes clumps among the carpet of wood anemones (*Anemone nemorosa*). We have moved it around and extended its range at various times. The scent of its flowers extends to the smell of the plant and its foliage. No wonder it is much picked.

The cowslip, *Primula veris* (20cm/8in), grows on open chalkland and limestone soils, in drier conditions than suit most primulas. The ploughing of old grassland and the use of chemical herbicides led to its decline in the wild but it has begun to return to unpsrayed verges and to colonize the banks of new roads. It does not grow wild around Dixter (the chalk downs are its nearest habitat). We introduced it to the garden where it maintains itself but does not increase, overall. Its flowers, in heads, are smaller and a deeper yellow than the primrose's and its scent is quite different. We also have red-flowered forms and a range of self-appointed hybrids, especially with polyanthus primroses, which my mother introduced to the drained upper moat at an early stage. It is amazing how persistent individual plants of that have been, through the years. They were probably used, initially, as spring bedding in the borders, then moved out.

The water saxifrage, *Darmera peltata*, grows in boggy places including wet meadows. Its bonnets of pink flowers are borne in spring before the leaves unfurl and they can make a good display, though the intensity of colouring varies considerably between different clones. With us its foliage often dies off brown, but in cooler climates, such as Scotland, where it thrives best, its foliage can assume brilliant autumn tints.

SAXIFRAGE

The meadow saxifrage (*Saxifraga granulata*), with clean white flowers on upright, hairy stems (50cm/20in) in early summer, grows in moist grassland on free-draining chalk soils. 'Granulata' refers to the knotted, egg-shaped bulbils that form where the lower leaf stalk joins the stems.

The water saxifrage, *Darmera peltata* (syn. *Peltiphyllum peltatum*), from the western states of the USA, thrives in damp spots. In spring it carries pretty heads of pink flowers on 38cm/15in stems, before the leaves emerge. These are peltate and subsequently rise to 1m/3¼ft, making a dense canopy of foliage.

ROSE FAMILY

Rosaceae is a mainly woody family. If you like to include shrubs in your meadow, there is the opportunity with some of the more robust roses. In the orchard at Sissinghurst in Kent, Vita Sackville-West festooned the original fruit trees with fragrant swags of rambling roses – mostly white, such as *R. filipes* 'Kiftsgate', 'Félicité Perpétue' and 'Madame Plantier' – but few survived being engulfed, both trees and roses being almost impossible to prune. I prefer to keep my meadows as meadow and I like my trees to be free of embrace. There are some of reasonable stature in what was once orchard and most of those are rosaceous – plums, a few apples, pear species, whitebeams and rowans (*Sorbus*), hawthorns, a couple of cherries. The grass being so competitive and my not proffering any manurial encouragement, those that flourish are not numerous but some do.

As to herbs, creeping cinquefoil, *Potentilla reptans*, runs through our meadows but is obvious only when it gets into the borders as a weed. Tormentil, *P. erecta*, with four- as opposed to the cinquefoil's five-petalled flowers, is common on short acid grassland on heaths, moors and roadsides. Silverweed, *P. anserina* (25cm/10in), is a pretty plant with its silvered, pinnate leaves. Despite its aggressively creeping habit, it has never made much headway, though present near the horse pond.

We have very little agrimony, *Agrimonia eupatoria* (40cm/16in), actually in the garden meadow areas,

An almost prostrate *Rosa gallica* in a Hungarian meadow, with single flowers of a clear deep pink shading to white at the centre.

though it abounds in nearby hedgebanks. A miserable thing, with spikes of tiny yellow flowers, followed by burr-like seeds that got into and clogged up our spaniels' ears (when my parents had them) and were the devil to comb out. *A. procera* is a larger, scarcer plant whose leaves and flowers have a scent of balsam – hence the common name of fragrant agrimony.

Meadowsweet, *Filipendula ulmaria*, always colonizes wet places, including meadows. Above pinnate foliage, it carries creamy-white, rather sickly scented flowering panicles quite late in the season, June–July. Its populations come and go according to rainfall and drainage. Although it can be aggressive, it is an endearing plant.

Meadowsweet, *Filipendula ulmaria* (1m/3¼ft), loves the damp bottom of the drained upper moat but its populations vary extraordinarily. First it became abundant, then almost disappeared, but is now back again in force. Its creamy, sickly-sweet-scented panicles are most welcome right at the end of the meadow's summer season, just before its first cut. It has an aggressive, rhizomatous rootstock, but snakeshead fritillaries cohabit with complete equanimity. In a damp area near the horse pond, the pink queen of the prairie, *F. rubra* 'Venusta', flowering in July, does well enough but everything is so lush in that area that you have to go to look for it. *F. purpurea* (80cm/32in) is one I have yet to introduce, but it makes a splendid show at The High Beeches, in central Sussex. The colour is intense magenta red. If you are impatient to get the grass cut in late June, these filipendulas are not for you.

Dropwort, *Filipendula vulgaris* (60cm/2ft), is a very different sort of plant, found wild on rather hungry chalk downs. Flowering in June, it has typical, cream-white meadowsweet panicles but held above a platform of tightly crimped, mossy-green foliage. As a cottage garden perennial, it is, I think, a nuisance, often too tall to sustain its own weight. Furthermore, its earth-coloured tubers are easily spread around and are hard to eradicate. Meadow conditions would seem to be ideal, but I have never kept it for more than a few years that way.

ELDERS

For rougher meadow areas, perhaps on a boundary or a shifting bank, where it will help to bind the soil, the herbaceous elder, *Sambucus ebulus* (*Caprifoliaceae*) is a good plant. Growing 1.5m/5ft high, it flowers in rather late summer; black elderberries follow. All needs to be cut right down at some point in its dormant season. This can be grown in sun or in shade and I have it along a wet ditch. It spreads quite rapidly by suckering, so you need to be prepared for that. There are other herbaceous species of elder with a similar habit, which might well be tried. Such is *S. adnata*, with strikingly bright orange berries.

WILLOWHERBS

Within the willowherb family (*Onagraceae*), there is dispute over the proper generic name for rosebay willowherb, which I shall here call by the familiar *Epilobium angustifolium* (1.5m/5ft). Within our context, it generally seeds itself where there is some promising bare ground and is not seen in dense turf. Often I do not want to be bothered by it, as it both runs and self-sows too freely. We not

infrequently resort to herbicide. But it is a beautiful thing, with spires of rich pink flowers followed by young seed pods which, in a sunny site, are scarcely less pink. The pure white *E. angustifolium* var. *album* (now called *Chamerion angustifolium* 'Album') has become popular in gardens, but will have no place in mine. It soon runs to seed and the dying plant is a shambles. As a runner it is no less aggressive than the type.

CRANESBILLS

The little species of *Geranium* that occur in all meadows need not detain us, but the meadow cranesbill, *G. pratense* (1m/3¼ft), is of the greatest importance, because it flowers quite late (June–July) and is a blue colour that we need. Besides which, it makes a show. My mother grew quantities of that from seed, and it is still in the areas where she planted it, besides self-seeding to a fair extent. We should really extend its range still further. The white version is also good and sometimes a plant turns up with some white petals, some blue. This cranesbill has to be cut down while still in full growth, albeit run to seed. This must reduce its vigour but not materially, and it bounces back with young greenery within a week or two.

The wood-cranesbill (80cm/32in), *Geranium sylvaticum*, is similar but the petals are white at their base. It is widely wild-distributed from Yorkshire northwards and is a woodlander in the southern part of its distribution, but more and more in the open as you go north in the British Isles. It is therefore a highly suitable meadow plant where summers are cool and also a candidate for the woodland edge (see Chapter 11).

The early-summer-flowering *Geranium clarkei* 'Kashmir White' is quite aggressive in a border and would, I thought, be ideal, given the check of grass, but it died out. The chalky-pink *G. endressii* has become well established on a steep bank which we used at one time as a tip, so everything there grows coarsely, but it seems to like that. I think it needs rather loose soil. I should love to get the magenta-flowered, dark-centred *G. psilostemon* growing enthusiastically and we have given it a good start in several places, but it only hangs on. There are many suitable geranium candidates for meadow treatment – the June-flowering bluish-purple *G. himalayense* (60cm/2ft), for instance.

The storksbill (*Erodium manescavii*), so called because it has even longer seed cases than the cranesbill, seems a strong enough grower to be a good meadow candidate, but that died out. I should try again. It has light magenta flowers over a long season.

UMBELLIFERS

The *Umbelliferae* are now *Apiaceae*, but anglicized words like umbelliferous and umbellifer will persist awhile. Most obvious of these, at Dixter, is cow parsley or Queen Anne's lace, *Anthriscus sylvestris* (1m/3¼ft), which flowers April–May – a mass of white froth. The leaves smell agreeable but the flowers are on the sickly side, which you discover if you cut them to bring indoors. In a poor area of meadow,

The tuberous-rooted water dropwort, *Oenanthe crocata*, is notoriously poisonous. It loves wet places and we sometimes spot-treat it with weedkiller to check its spread. But its flowers in late spring are purest white umbellifers and are odourless, so I find them suitable in large flower arrangements – far better for this purpose than cow parsley, *Anthriscus sylvestris*.

cow parsley is greatly reduced in vigour and offers little challenge to other flora, but in stiff, moisture-rich loam, it is apt to take over.

A little later and far better for cutting for the house comes the water dropwort, or water hemlock, *Oenanthe crocata* (1.2m/4ft). It is a strongly branching plant with shiny leaves and whiter flowers than cow parsley and it is odourless. This thrives in wet places, where it seeds aggressively and competitors are at a considerable disadvantage. It looks nice with yellow flag irises (*Iris pseudacorus*). The roots are tuberous and the plant is poisonous (so is hemlock, as Socrates discovered). For chalky grassland there is the parsley-like unbellifer, burnet-saxifrage (*Pimpinella saxifraga*), and wild parsnip (*Pastinaca sativa*), which forms clouds of yellow when seen from a distance.

Pignut, *Conopodium majus* (40cm/16in), is a very refined umbellifer, with finely divided leaves and light heads of white flowers, May–June. Its dark brown tubers won't stand much competition and it has never taken to my meadow areas but is dominant in some I know, one of them only a few miles away, and a delight in its season.

Astrantia (50cm/20in) are distinctive on account of the ruff of conspicuous bracts and the pincushion effect of the flowers within. Any of them would be worth trying in a meadow and, as they self-sow freely in the garden, there should be no problem in finding stock. But I have yet to get round to doing anything about this. A friend in Scotland is having encouraging results.

Our native hogweed, *Heracleum sphondylium* (1.5m/5ft), is beautiful on roadside banks, where its colour variation, often pink, is distinctive as are its fairly large umbels and its latish, July season. But its presence is usually a sign of disturbed ground which, in my meadows, means where badgers have dug themselves a colony of little pits in which to defecate. This completely alters and coarsens the flora, ousting smaller species. Which is not the fault of the hogweed, of course. It is usually biennial and a fine, upstanding plant, not notably prone to causing skin allergies (although it can), as does its reviled yet stately cousin, the giant hogweed, *H. mantegazzianum*.

Wild carrot, *Daucus carota* (90cm/3ft) has pale pink umbels before opening to white. It looks good in Keith Wiley's Cretan meadow at the Garden House in Devon. In the USA it is classified as a pest.

Astrantias thrive in one of the Hungarian meadows I visited. In the garden I have a slight antipathy towards *Astrantia major* as it self-sows freely and most of the progeny is inferior. However, in a meadow context, self-sowing is a plus point. The dark red colour variants, such as 'Ruby Wedding', which are often lost in a border, show up well in a meadow.

GENTIAN FAMILY

Common centaury, *Centaurium erythraea* (30cm/1ft), belongs to *Gentianaceae*. It is a biennial and its rosette of leaves in the first year has always reminded me of *Gentiana acaulis*. It is one of those heart-warming plants which sulks in dull weather and closes at night, but opens into a wealth of little pink stars on a generously branching inflorescence when the sun shines, any time from summer to autumn. It is always in my garden, somewhere, usually in paving cracks in our sunk garden or in one of the less disturbed borders above this. It is also a regular in our woods, after coppicing. In most of our meadows, life is too competitive, but if the turf has been skinned, for some reason, it may take advantage of the gap. Sometimes you see it in a great colony (I'm thinking of a meadow in West Sussex) and you catch your breath.

Thin turf is also the place for *Gentianella amarella*, the autumn gentian or felwort, a late-flowering biennial with purple flowers.

COMFREYS

Boraginaceae gives us the common comfrey, *Symphytum officinale* (1m/3¼ft), a coarse plant, with large hairy leaves that is popular with bees. Its colouring and flower size vary greatly. Typical of the vegetation in a wet meadow, we saw it in such a situation in Hungary. At Dixter it has always grown around our principal rubbish heap and in a wet hollow that was once a pond, but drained. It is an unexciting plant because its tubular flowers are such a murky shade of purple. However, from a Romanian (Transylvanian) ditch, some twenty years ago, I introduced a clone with rich red flowers, and that has adapted well.

I should love to get *Symphytum caucasicum* (80cm/32in) going in a damp meadow spot. Having failed once, I am not sure which way to turn. It is the bluest of all the comfreys but inconveniently invasive in most border sites. Currently we have it to one side of a holm oak hedge, on the divide between garden proper and meadow. It is happy, but might be too dry in the long run.

LABIATES

Lamiaceae are poorly represented in my meadows. Selfheal, *Prunella vulgaris*, is a principal weed in our lawns (what's left of them), withstanding close mowing with equanimity, but it is a dull thing, with its purplish flowers. The other lawn interloper in one rather damp, shaded area where we don't use herbicide is the creeping Corsican mint, *Mentha requienii*, which gives off a delicious minty aroma when bruised. Its leaves are tiny. In shade you scarcely notice its flowers, but they can be abundant little mauve things out in the sun where, however, the plant looks less healthy.

Selfheal, *Prunella vulgaris*, has a creeping habit and thrives on mowing. Its purplish-blue flowers make no great show.

Other labiates worth including in a meadow are betony (*Stachys officinalis*), bugle (*Ajuga reptans*), wild basil (*Clinopodium vulgare*) and wild clary (*Salvia verbenaca*). The North American obedient plant (*Physostegia virginiana*) needs moisture and heat to produce its spiky flowers in late summer.

PLANTAINS

In the *Plantaginaceae*, there are two prominent plantains. The greater plantain, *Plantago major* (prostrate) is a familiar lawn weed, making ground-hugging rosettes of broad, ribbed leaves. It does not cope at all with long meadow grass, which is where the ribwort plantain, *P. lanceolata* (up to 50cm/20in), comes into its own. It has narrow, lanceolate leaves and the flower heads are black, peppered all over with white anthers – quite a handsome arrangement, when you get as far as actually looking at it instead of taking it for granted. It is abundant. I tried, without success, to introduce the hoary plantain, *P. media*, from Hungary, where the strain in question had quite a strikingly pink inflorescence. In England it is common on grassland on chalky or neutral soils.

FIGWORTS

Scrophulariaceae are a fascinatingly variable lot. Here belong antirrhinums and foxgloves, not to mention common figwort (*Scrophularia nodosa*) itself, so named because the globular red flowerbuds and root protuberances were thought to resemble figs, an old word for piles. Several *Veronica* inhabit meadow turf but the most noticeable is an exotic, originally from Turkey, the slender speedwell, *V. filiformis*. It is mat-forming and its bright blue colouring has special charm in its spring season. However, it is terribly invasive. It has come to me from my neighbours and I do not want it, so we treat with herbicide when we see it, but victory will never be complete. Not only does it invade lawns but it inevitably moves from them into borders. Perhaps I am being neurotic about it, but I like my weeds to be invited.

Among meadow grasses, the tall-growing ribwort plantain, *Plantago lanceolata*, is quite pleasing in its quiet way.

There is a range of scrophs, as we know them, that are semi-parasitic. In part they fend for themselves, having green leaves for this purpose, but they also parasitize a range of meadow turf hosts. *Euphrasia*, called eyebright, is one of these – a complex genus of low-growing annual plants found in short grassland, the aggregate being known as *E. officinalis*. Only a few centimetres high, as a rule, it has small white flowers mottled with purple and with a yellow eye, not unlike the colours of a bruised eye. Even so it is a sweet little thing when looked at closely, but easily overlooked.

Annual hay rattle, *Rhinanthus minor* (30cm/1ft), which I have already described as a great tool for reducing vigour in a meadow sward, is the much showier, yellow-flowered parasite that we have introduced. It also parasitizes other plants, but not so seriously as to be a worry. The ripe seeds do rattle inside the brown calyces that are shaped like purses or seashells. Lousewort, *Pedicularis sylvatica*, with pinkish-purple flowers, does the same job as hay rattle, but on acid grassland.

BEDSTRAW FAMILY

The *Rubiaceae* includes calcifuge heath bedstraw, *Galium saxatile*, which I have seen making low sheets of tiny white flowers, cheek by jowl with the yellow lady's bedstraw, *Galium verum*, which needs well-drained neutral or alkaline soil. The latter is also short, but the spikes are so dense with flowers that it creates sheets of colour in grassland from early summer to autumn.

TOOTHWORTS

In the *Orobanchaceae*, we have a family of fully parasitic plants, making no green leaves at all. The one I have introduced (it comes from southern Europe) is a toothwort, *Lathraea clandestina*, given to me as a lump of roots dug up from where it was already established. I simply planted it, in two or three places at the foot of willows (which, among other hosts, it fancies), and it flowered from the very next year. Close to ground level, it makes clusters of bright purple, hooded flowers. Along the Backs, at Cambridge, it makes a terrific show in spring and at a distance you might mistake it for crocuses. It has never made such a display for me but I love to see it, near to the horse pond (also in my long border). It has a long flowering season and seems capable of straying from the original host and getting along among other things.

CAMPANULAS

I find most campanulas (*Campanulaceae*) difficult in my meadows, though I am still trying to establish the harebell (bluebell in America), *Campanula rotundifolia* (30cm/1ft), in the topiary lawn. But it is none too easily pleased. I have raised several batches from seed and put out strong transplants.

I have seen the tall *Campanula lactiflora* (1.5m/5ft) stunningly successful in Scotland but it seems to find the competition too great in my turf. Creeping bellflower, *C. rapunculoides* (80cm/33in), should be kept out of the garden proper at all costs. It spreads uncontrollably both by making rhizomes, vegetatively, and by self-sowing. Turf, however, checks its progress perfectly and it certainly is pretty.

The parasitic purple toothwort, *Lathraea clandestina*, makes fang-like tufts in damp spots at the foot of willows near the horse pond at Dixter.

8. Bulbs & Corms

A MEADOW WITHOUT BULBS would be unthinkable to many people. In fact, it is perfectly possible, allowing a different cutting regime. But with bulbs you have to take into account that some of them, including narcissi, will be showing through before the new year, which means that you dare not mow after November. At least, we dare not and the only area where bulbs have not been planted at Dixter is the one-time putting lawn in the topiary garden, which has only recently become meadow. I've already described our planting method (see Chapter 4).

SNOWDROPS

From the first, that is around 1912, snowdrops were planted in our orchard, but only *Galanthus nivalis* was considered. Recently, I have tried the excellent cultivar 'S. Arnott', additionally. In borders the bulbs quickly become overcrowded, pushing each other out of the ground and, unless thinned, killing themselves out. This does not happen so quickly, if at all (but we shall see), under meadow conditions.

However, snowdrops do not thrive in the densest turf, or so I have found. Which means that they are little use in the most open areas, but need some shade, where the turf is thinner. I have pretty well got the measure, by now, of where these areas are. You cannot have too many snowdrops.

This piece of meadow, either side of the front path at Dixter, looks its most colourful and lively in early spring when crocuses and our native daffodil, *Narcissus pseudonarcissus*, called Lent lily, are flowering. They all increase by self-sowing. We allow no coarse, tall-growing bulbs in this area, which is lively in autumn with another species of crocus, the near-blue *C. speciosa*.

They respond to changes of temperature by opening and closing, yet are never as sulky as crocuses can be. They dance with air movements; they are most of them deliciously scented and they are besieged for nectar by hungry bees. But any galanthophile would think me woefully short of them. Well, let them think. I don't want to join the galanthobores.

CROCUSES

Actually, I am far more passionate about crocuses than snowdrops, and we are still planting them, with our old-fashioned bulb planter, taking out single plugs and planting them singly, by the thousand, each plug hole receiving a handful of old potting soil at the bottom to give the bulb a good start. I have Fergus right behind me there. There is nowhere that you cannot plant crocuses, provided they can happily cope with turf in the first place. Originally the Dutch yellows and *Crocus vernus* cultivars were the ones planted, between the areas of narcissi, in the orchard. The Dutchies are sterile, but build into clumps, which you can divide, after flowering, and spread around. We have done some of that (not enough).

The *Crocus vernus* types, in shades of white, mauve, purple, often striped, seed themselves abundantly. Give them a start and they will do a lot for you in that area, and they are as good among the daffodil clumps as out on their own, because their flowering, centred on early March, comes before the daffodil shoots are tall enough to mask the display. And what a display! On a sunny day, it is breathtaking, and we have sheets of them, but you have to catch them right. No flower is moodier. If you are only at home by day at the weekends, a whole crocus season may slip by without your seeing them fully relaxed and open at all. You'll only hear about how wonderful they were the moment your back was turned.

I started to add *Crocus tommasinianus* in the late 1940s. That is quite petite and refined, generally at its best in February, though often, in the early springs we have been enjoying, well out in January. It increases by self-sowing and has the refinement of a true species, but is almost invisible when closed, being likened to a toothpick, and only showing its cheerful mauve interior when the weather warms up. It is worth buying 'Whitewell Purple' for its extra richness of colour. It spreads well, both by seeding and by clumping up.

Of similar habit and flowering season, so that it makes a good contrast with and companion for *Crocus tommasinianus*, is *C. flavus* subsp. *flavus* (formerly known as *C. aureus*). It is the chief progenitor of the 'Dutch Yellow' but smaller, a much richer shade of true orange and fertile, so that it can self-sow. It is not so cheap to buy as the Tommies.

Another early crocus which I like to grow in quantity is *C. chrysanthus*, which has many named cultivars in many colours, but you also get exciting colour variants from your own stock, where it is cross-breeding and self-sowing. Some of our densest areas are next to the front path, along which one is often passing so that it is easy to see what's going on. A favourite cultivar (when correctly named, which it frequently is not) is 'Snow Bunting'. It should be one of the earliest in flower, by the end of January, and it clumps up well. The colour is cream-white (not pure

RIGHT *Crocus tommasinianus*, of which this is a coarsened, bastardized version, flowers in February and is a great self-sower. The closed blooms, in dull or cold weather, have been likened to toothpicks and are almost invisible. When the sun shines and when the wind turns to the south, they open wide and make a great show, while always remaining elegant.

BELOW *Crocus chrysanthus* is a variable species of which many named varieties have been selected, not least by E. A. Bowles of Myddleton House in Enfield, London, who named them after birds. This one is 'Snow Bunting', one of the first in flower in January. It clumps up well and the flowers are deliciously honey-scented. It is my favourite.

white) and the blooms are apt to be a bit ragged at the margins. They are dark-striped on the exterior. When open, they have a delicious scent, much like that of some early tulips.

We are planting lots more *Crocus chrysanthus* cultivars in the orchard. The typical flower shape is waisted, rather like an electric light bulb. We have others of a similar flowering time, like *C. sieberi* 'Violet Queen'. But some, like *C. imperati*, I have found no use in turf at all. It is in any newly introduced case advisable to work up stock in a border, where multiplication is rapid, and then to transfer what's surplus to turf.

Autumn-flowering crocuses are another subject altogether, yet almost as important, at Dixter. Those which make their 'grass' (a frequent designation of crocus foliage) at the same time as flowering are not suitable in turf that you want to give a final cut in late autumn. The regime with autumn crocuses is to cut by late August or early September. Sometimes you can fit in two cuts before the crocuses flower and have the background turf really short for them. The autumn-flowerers are very responsive to water availability. After a heavy rain in the first half of September, they are liable to be up and flowering only days later. If you are clever, you may

Crocus speciosus is as near to blue as this flower can get, and it is as sweetly scented as 'Snow Bunting', but autumn flowering. The flowers have deep blue stripes which contrast with the brilliant scarlet stigmas. Rather weak-stemmed, the flowers lie on their sides after opening, but that doesn't spoil them. Its foliage, like bright green 'grass', stands out in its freshness in February, before grass itself has greened up much. Spring is a good time for spreading this crocus, and *C. nudiflorus*, around.

be able to fit in a cut immediately after the rain and before the crocuses have responded. If there is drought throughout the period, it may delay the crocuses' flowering by weeks. Even though the climate in Northumberland is much cooler than with me, I have noticed *C. nudiflorus* flowering several weeks before mine, simply because they are damper.

The best species of autumn crocus for my purpose, ones which do not make their 'grass' till the new year or later, are *C. speciosus* and *C. nudiflorus*. I have never had much success with *C. kotschyanus*, having bought a strain which clumps up with a will, but scarcely ever flowers. But it can be free and is then the earliest autumn-flowerer to bloom. *C. speciosus* is the only reasonably cheap autumn-flowerer to buy, though you can soon work up stock under border conditions. The heavily striped flowers make as blue an impression as I know in any crocus and the scarlet stigmas are in brilliant contrast. The 'stem' (not a true stem) is long and rather weak, so the bloom habitually lies on its side after the first day or two. I don't mind, as it continues to flower well from a recumbent position. It is sweetly scented. My clone never makes seed but it spreads into dense patches and spring-flowering crocuses will not grow in these areas. The flowering time of *C. speciosus* can be manipulated to an extent. I have a patch of it beneath the outer branches of my favourite hawthorn, *Crataegus orientalis* (syn. *C. laciniata*). This carries crops of large orange haws in September and their weight brings them down to near ground level where the crocuses are. If these are still dormant, a heavy drenching will get them started so that they will perform simultaneously.

Crocus nudiflorus is my other winner. There, again, it does not make seed but it has a rhizomatous habit, barely troubling to make true corms, so it cannot easily be dried off for handling by the bulb merchants. Once you have it, it will make large patches and is easily spread around in the spring, when 'in the green', as the expression is. Its 'stem' is reasonably strong, and the habit upright, in colouring a rich shade of light purple. My large colony is in the upper moat. Quite a sight in late September. My original clump was given to me by the owner of a Northern Irish garden, in the early 1960s, and I have gone on from there with some persistence.

When these autumn crocuses have finished, say in late October, the meadow grass can be given its final cut. Colchicums behave in much the same way, though some of them are in flower by mid-August. As the common name meadow saffron suggests, they naturally inhabit undisturbed meadows (not many of which still exist), but they have never grown for me with wild abandon and do not self-sow, although some of them seed. Still, I have quite a lot of different kinds which maintain themselves. Among my oldest clumps is *Colchicum agrippinum* whose narrow segments are checkered. The double-flowered kinds, which flop rather conspicuously in a border, are sufficiently supported by grass. All the coloured colchicums have a pinkish flush to their mauve colouring, which you never find in crocuses. The obvious botanical difference is that crocuses have three stamens, colchicums six. All are lumped together by our more ignorant public as autumn crocuses. In our borders, colchicums make up and become overcrowded quite quickly, so there is the frequent opportunity of transferring them to meadow.

NARCISSUS

Now we come (or return, by the calendar) to *Narcissus*. The daffodil orchard is expected to be at its peak around 12 April, but varies wildly from year to year. The daffodils, planted around 1912, were cultivars fashionable at that time. If I see the same varieties in another garden, I know that it, too, was planted up in the early twentieth century.

You may think that mixtures give best value for money, but they confuse the eye and those which flowered early will spoil the effect of those flowering late. I have touched on the grouping of different kinds in our orchards, as Robinson advised. The first to flower, 'Princeps', is followed by the mainstream 'Emperor' and 'Minnie Hume', a white narcissus mixed in with a smaller white one, 'Lillie Langtry' (or it might have been 'Mrs Langtry'), named after King Edward VII's favourite, who was popularly known as the Jersey Lily. A little later comes 'Conspicuus', sold to us as 'Barrii Conspicuus'. It is yellow, stained orange with a small cup rimmed with bright scarlet-orange and has a lovely scent. Latest of all is the poet's narcissus or pheasant's eye, *Narcissus poeticus* var. *recurvus*, usually out in early May and deliciously scented. How freely it flowers varies greatly from year to year and I have, from long observation, concluded that if the weather turns unduly

Our orchard was planted up with daffodils and narcissi such as were fashionable at the time, around 1912. This was before the coarse modern whoppers had been bred. Plenty of space was left between groups. They don't bring on indigestion but allow smaller things – like snakeshead fritillaries, crocuses and wild orchids – to play an important role.

wet in May, so that fungal disease prematurely destroys the narcissus leaves, the following year will be a virtual blank.

How can you be sure that narcissi will naturalize successfully for you? Robinson insisted that if the soil was light and dry, they would be no use, and I think he was right. My soil is heavy and for many years there were few problems, but then came some winters that were so wet as to prevent adequate drainage, and that killed many narcissi that lay in water. They never properly recovered in those areas. The orchard is in ridges and furrows, which is a relic of the way it was ploughed in medieval times. On the ridges there are no problems.

In other parts of the garden we have eschewed the big chaps. My mother raised from seed quantities of our native Lent lily, *N. pseudonarcissus*, which flowers in March, obtaining her original stock from a friend, Molly Liddell, living at Place House, Peasmarsh, about 8 km/5 miles away. Whether that stock was wild, I do not know, but certainly there have long been remnants of the species in the area. So my mother, having raised her stock in seed boxes, then planted it out, mainly in the front path meadow, where all narcissi are small. They have done extremely well, and in other areas also, self-sowing. This is a little yellow trumpet daffodil. It has a wide distribution as a wilding through England and Wales but is now quite localized and rare.

Similarly a small yellow trumpet but about half the size is *Narcissus minor*, from northern Spain, my original stock of

The yellow trumpet daffodil 'Emperor' flowers mid-season, in early April, and is always reliable for a fortnight's display, but not for any longer than that. It was a fashionable cultivar in the years preceding the First World War. Gardens where it is seen can be dated to that period.

LEFT The poet's narcissus, *Narcissus poeticus* var. *recurvus*, is otherwise known as pheasant's eye, having a very small cup with a thin red rim. It makes big clumps but quite frequently goes in for a non-flowering year. When it does flower it is one of the last, in early May, and is deliciously scented on the air. By the end of June it is making new roots so you need to plant the bulbs as early as you can. The earlier the better with the whole narcissus tribe, in fact, making the end of September your deadline (if you live in the northern hemisphere that is).

BELOW Where you want to keep your meadow contents small and refined, these three daffodils will serve you well in short turf.
TOP *Narcissus minor* is like a smaller version of *N. pseudonarcissus* and flowers in March.
MIDDLE *N. asturiensis* is smaller still and earlier.
BOTTOM *N. bulbocodium* var. *citrinus* is the lemon yellow, and infinitely superior, variant of the hoop petticoat daffodil, which has very small segments but a much inflated, lampshade-like corona. It self-sows and increases freely, flowering over a long period, January to early April. Each bloom lasts for three weeks.

which I obtained from the same Northern Irish garden as *Crocus nudiflorus*. Small though it is, *N. minor* copes extremely well with rough grass; even so it is best to work up stock quickly in a border, where it makes clumps that can be divided. I am growing this in several rough areas where there is no competition from larger species and hybrids, the most successful from a scenic point of view being among a colony of dog's-tooth violet, *Erythronium dens-canis*.

An even earlier and even wee-er yellow trumpet species is *Narcissus asturiensis*, which is growing in unpropitiously poor ground beneath an oak, but is perfectly at home (though beloved of slugs). Its first flowers are nearly always out well before the end of January.

There are other species I have tried with or without great success. My strain of the hoop petticoat daffodil, *Narcissus bulbocodium*, is sterile and it has never flourished in dense turf. As I have said, its bright yellow colouring is exactly that of dandelions, and as the two flower simultaneously, you could scarcely distinguish the daffodil until evening, when the dandelion flowers closed. Far more successful have been my fairly recent experiments with the pale yellow variant, *N. b.* var. *citrinus*, which sets seed and self-sows. I have it in very poor soil under an oak, where the turf is thin. But it is happy, the first bloom out in January, continuing into April, and

each bloom is long-lasting. From modest beginnings, I am gradually getting a nice colony. But I have not succeeded with *N. cyclamineus*, which makes huge colonies in the Savill Garden, Windsor Great Park in Surrey, and elsewhere. Obviously, I have not been trying seriously enough, but we have recently made a fresh start.

SUMMER SNOWFLAKES

Leucojum aestivum, the so-called summer snowflake (it flowers in spring), has very much the habit of a daffodil with similar foliage. My mother was given it by a neighbour and planted it in the upper moat, where it flourishes to this day. The leaves, which push through in November, are very dark green, not at all glaucous. The first blooms are often out before the end of February and at their best by late March. After that, the weather tends to lay both foliage and flowers. The latter are in an umbel of five or so bell-shaped blooms, opening sequentially. They are white with a green tip to each segment. It grows as vigorously in grass as in a border.

Leucojum aestivum has much the same habit and foliage as an ordinary daffodil, but white bell flowers with green tips, four or five of them to a stem, opening in succession. Although called summer snowflakes, they are often at their best in March, apt to get flattened by wind and rain. 'Gravetye Giant' is sturdier and to be preferred, I consider, flowering in May.

It was *Leucojum aestivum* subsp. *pulchellum* that my mother acquired and which is most commonly seen in gardens and it hails from the Mediterranean. (According to Gerard, 'These plants do grow wilde in Italy and the places adiacent, notwithstanding our London gardens haue taken posssession of them all, many yeeres past.' That was at the end of the sixteenth century.) However, subsp. *aestivum* is native to Britain, and known as the Lodden lily, the Lodden being a tributary of the Thames in the Twyford area. There it grows in river meadows. I am not certain of my facts, here, but I believe this to be identical with 'Gravetye Giant', which I grow in a frequently waterlogged area among dogwoods (*Cornus alba*) by the horse pond. It is a sturdier plant that does not get laid and its bells are larger and showier. I believe it would grow actually in shallow water, but whenever I have attempted this, the bulbs have floated away! The main point here is that both subspecies of this leucojum will flourish in dense turf, waterlogged or well drained. Where I grow it in a fairly exposed position, 'Gravetye Giant' flowers into early May.

Some strains of *Fritillaria meleagris* include albinos and there is a whole range of intermediate shades. *F. m.* var. *unicolor* subvar. *alba* is the one to order to get pure white, if that's your preference.

FRITILLARIES

Another wet-meadow-loving bulb, though it does not fancy prolonged flooding, is the snakeshead fritillary, *Fritillaria meleagris*. It flourishes in any stiff soil, so long as not too acid. The most spectacular display is in the damp meadow at Magdalen College in Oxford – a purple haze covering the north-eastern half of the meadow in late April. On the Continent it can be seen around the flood plain of the Rhine. Bulbs are cheap, so you can easily, with a little patience, establish a colony. Where my mother obtained her first bulbs, I forget, but she was assiduous in raising them from seed, planting them out when large enough, as she did *Narcissus pseudonarcissus*. Fritillaries were among her principal contributions to the Botticelli garden, but she also established colonies in the orchard, either side of the front path and elsewhere. The flower is usually a single (there are sometimes two) bell, rich purple and heavily checkered. However, we seem to have even more albinos than of the purple and there are many intermediate shades. Flowering starts well before the end of March and continues through April. One of the best bulbs for naturalizing, as it self-sows freely, it is caviar to pheasants, should you be plagued with these birds.

Snakeshead fritillaries, *Fritillaria meleagris*, are considered not to be true natives to Britain but they have colonized water meadows, as here at Magdalen College, Oxford. The naturalist W. H. Hudson describes a field which, at a distance, he thought to be ploughed but discovered, on approaching, to be a mass of the 'checkered daffodil', as he called it. They do indeed naturalize easily, especially on heavy soils.

TULIPS

At Highgrove, in Gloucestershire, the Prince of Wales has planted massed tulips in grass to create swathes of purple and scarlet to either side of a closely mown path. I should love to be able to say that I am successful with tulips in turf, which makes such a flattering setting for them, but I am not and I know of no one who is. Always, they have to top up with fresh bulbs if a regular display is to be enjoyed. That is a perfectly feasible procedure, done on a regular, planned basis, but, as with daffodils, keep the same varieties to the same areas and don't allow an unplanned mixture. If there are surplus bulbs to be disposed of, rather than throw them away, they should certainly be planted in a meadow. The greatly reduced size of their blooms is a positive asset, especially in the large, normally overfed hybrids.

Some tulip species with a running, rhizomatous rootstock will thrive in a meadow, but never or hardly ever flower. *Tulipa sylvestris* is one such and some people do flower it (it has got into the floras), yellow and deliciously scented. I know of a garden near Nottingham where a sizeable colony in grass is really prolific in some years. But it is only vegetative at Dixter, as is *T. orphanidea*, which is a bricky-red Turkish species that has flowered for me once in more than sixty years, although its leaves appear regularly. Full sun is necessary, of course, and if you secured a free-flowering strain that adapted to our cool climate, you might well get results. I have to chalk up the late-flowering, rich red *T. sprengeri* as another meadow failure, although it self-sows and spreads enthusiastically in my borders, always flowering well. It has been known to perform in turf, so we will try again in a number of different places in our meadow.

Species tulips are often shy-flowering but spread vegetatively to form colonies in grass. **FAR LEFT** The sweetly-scented yellow *Tulipa sylvestris* in a meadow at Holme Pierrepoint Hall in Nottingham. **LEFT** Mauve and yellow *T. saxatalis* Bakeri Group growing in a Dixter meadow.

DOG'S-TOOTH VIOLETS

The European dog's-tooth violet, *Erythronium dens-canis*, was a failure and died out when my mother tried it but has been a great success since I gave it another bash, starting with a clump 'in the green', some forty years ago. It does best in fairly open-textured turf, partially shaded by trees. It appears, suddenly, in early March and may be flowering a few days later.

There are two smooth-textured leaves to every bulb and they are beautifully marbled in purple and green. I should add, from recent experience, that an inferior strain is being widely marketed with less boldly marked foliage and with wan pink flowers. The turkscap flowers are pendent, opening out with reflexed segments when it warms up a bit or the sun is on them. They are a lively shade of mauvy-pink. You must make a point of going to gloat over them daily, as the display is brief and may be over in a week. There is only a little self-seeding. To get a good colony, you really need to split clumps and replant singleton bulbs soon after flowering. We have done this quite extensively, especially on the once-quarried bank that is set back from the horse pond. Ideal companions are *Narcissus minor* and primroses. Sometimes clumps of this erythronium are largely blind with a lot of leaf and few flowers. This suggests that they have become congested and need dividing.

One of the best species for making a colony, because it has a running rootstock, is the American *E. revolutum*, impressive in the Savill Garden in Windsor Great Park. It is showiest in its pink-flowered form. I am growing this at the other end of the above-mentioned bank, where the turf is very thin, and it is making progress, although the experiment is still quite young. It flowers a whole month later than

The European dog's-tooth violet, *Erythronium dens-canis*, flowers in March, suddenly appearing as from nowhere.
RIGHT The flowers open wide to a turkscap when warmed by sunshine. They like dappled shade and naturalize nicely with *Narcissus minor*.
FAR RIGHT The beautifully marbled leaf of a seedling.

E. dens-canis. Other species and hybrids should be tried. There is endless scope for experiment. But always avoid dense turf. This nearly always means that there will be some overhead shade. You will get a feeling for the sort of area a plant would like.

ALLIUMS

The turf at Dixter is too dense to suit the nodding onion, *Allium cernuum*, a clump former that is native to meadows and prairies in a variety of soils in the USA.

The genus *Allium* includes some terrifying weeds but they can sometimes be given a safe outlet in turf, which restricts their takeover ambitions. One of the worst, which I should on no account wittingly introduce to my garden, I have already mentioned as looking charming in the West Country. It is the beguiling *A. triquetrum* (30cm/1ft) from the western Mediterranean, easily identified by its scapes which are triangular in section and give rise to the common name, three-cornered garlic. The flowers, in April–May, are white. But it is such a spreader, both from seed and bulbils, as often to defeat control. My one-time neighbours introduced it into the turf on our boundary. It invaded us like lightning and we are still trying to rid ourselves of it. I suspect that *A. paradoxum* (few-flowered garlic), an introduction from the Caucasus, *A. oleraceum* (field garlic) and *A. vineale* (crow garlic), both with greenish-white to pink flowers on wiry stems, might be similarly invasive.

Allium roseum, rosy garlic, another stinker from the Mediterranean, is a charmer as I have seen it in both the Cambridge Botanic Garden and at Kew, naturalized in meadow, with heads of soft pink flowers on 50cm/20in stems. Most clones make bulbils within the flower head, so it is apt to spread around aggressively, but I should like to try it. *A. sphaerocephalon*, the round-headed leek, which I have seen in meadows in Hungary, is another possibility. I have it in my borders and it is not at all aggressive, but I have not got around to trying it in turf. With its elliptical, purple flower heads, it does not flower till July (at 50cm/20in, the stems flexible), and so, in meadow conditions, it would need siting where the first cut was not till late September. It grows on the dry limestone crags in the Avon Gorge with other southern European onions, including rosy garlic and honey garlic (*Nectaroscordum siculum*).

All the aforementioned onions form bulbs. I now come to a clump-former, *Allium cernuum* (30cm/1ft), with generous clusters of pinky-mauve-purple flowers, on curved pedicels, which develop into small, rounded seed capsules. Nice with old-fashioned laced pinks in the border, I have yet to try it in a meadow situation. It grows in mesic prairies and is especially common around southern Lake Michigan.

STAR-OF-BETHLEHEM

The star-of-Bethlehem (one of them, anyway), *Ornithogalum umbellatum*, can be quite a nuisance under border conditions, because of the numerous bulbils sired by each bulb, but it is ideal meadow material, its white, green-striped stars showing up well even where the grasses around it are twice its own height (20cm/8in). They open and become visible only when the sun shines, being borne in corymbs, most attractively presented. April is its season.

Ornithogalum nutans, with racemes of green stars, is a dead loss under meadow conditions, becoming effectually invisible in a green surround. It will grow well enough but I believe visibility to be a more important element in meadow gardening than the number of species successfully grown.

GRAPE HYACINTHS

My mother planted a clump of spring-flowering *Muscari armeniacum* in her Botticelli garden and it held its own for many years, its pure blue colouring being a great asset in this otherwise non-blue context. It eventually died out and my later attempts to establish it have not succeeded, though I feel it should be possible. *M. neglectum* (syn. *M. racemosum*) with blue pearl-flowers, a speciality of chalky verges in Cambridgeshire and sandy areas in Suffolk, could be worth trying on light soils.

BLUEBELLS

First *Scilla*, then *Endymion*, currently *Hyacinthoides*, what the British call bluebells have suffered more than most genera at the hands of the botanists. (My American friends use this common name as an alternative to harebell, *Campanula rotundifolia*.) Our native bluebell is *Hyacinthoides non-scripta*. It has a one-sided raceme, drooping at the tip, and the segments curl back at their tips. It is strongly sweet-scented, very much so on the air, when growing in quantity. In my part

I do not recommend introducing *Hyacinthoides non-scripta* to your flower borders – the bluebell seeds freely and the bulbs go deep – but in damp meadows it makes welcome patches of blue in spring and early summer and later the dying foliage is well camouflaged.

of Britain, it is a woodland flower, though most prolific after the wood has been coppiced and is letting in more light. In the garden, it seems to grow as well in sun or shade. The bulbs go deep. It is hardly an important meadow ingredient at Dixter, but makes welcome patches in April and May of a more intense blue colouring than the Spanish bluebell. This, *H. hispanica*, is the species more commonly grown in gardens. It has an erect raceme, with flowers all around it, and they are bell-shaped, the segments not recurved. There is little scent. Typically, it is pale blue and there is much of it in the upper moat, where it contrasts nicely in its April – May season with the rich colouring of early-purples, *Orchis mascula*. There is plenty of this bluebell in the orchard, too, where there are also pink- and white-flowered variants. The Spanish bluebell is strong on these. Compared with our native bluebell, it is coarse, but welcome in many places. In turf you do not notice any dishevelled appearance after flowering, as is so unpleasantly obvious in many London gardens.

CAMASSIAS

From bluebells there is a natural progression to the North American camassias, which closely resemble bluebells in their habit but have an upright raceme of open, star-shaped, blue flowers, replacing bells. They mostly come from the western USA, in mountainous regions overlooking the Pacific. They are often under snow in winter but then benefit from snowmelt in what, to me in Britain, would be late spring. My diary tells me that the one species that I saw in quantity (I was never sure which species; my companion thought *Camassia leichtlinii*, but the colouring seemed unusually deep blue for that) was on 27 May 1986. We had just crossed the Carson Pass (still under snow) on the Sierra Nevada, between California and Nevada itself. The ground was soaking wet from snowmelt, but just coming into bloom was this large area of camassias.

Some of the larger hybrids and species selections are sterile. To increase them, you need to divide the clumps and spread them around. Most of ours grow in the areas either side of the front path (see page 73) and they arouse much interested comment.

The one which not only clumps up but also sows itself around is *Camassia quamash* (syn. *C. esculenta*). It has edible bulbs and is part of the native North Americans' diet. The very word quamash suggests scrunching, to me. These used to be very cheap to buy and I did so on an annual basis. I found that I was getting different strains, some earlier-, some later-flowering, some a fairly pale blue but most an intense blue. Their flowering season is from late April through most of May, and they are not unduly tall – about 35cm/14in. They do not always flower abundantly; sometimes there is more of their grass-like foliage than of bloom, but on the whole they give a good account of themselves and they always set plenty of seed. This is not ripe till late July and I watch the capsules vigilantly, so as not to mow the grass before they have scattered their contents.

Almost twice its height is *Camassia cusickii*, which is showy and not so pale a blue as to be ineffective (though difficult to photograph). In its cultivated version this is sterile, as also is *C. leichtlinii* subsp. *suksdorfii* 'Electra', with large, pale blue stars. We transfer surplus stock from the borders into the meadow. Its season is short. We are also, with some success, growing the creamy-white camassia *C. leichtlinii*. In the borders, this is floppy, but grass supports it and it gleams out most effectively. I have yet to try *C. scilloides* and *C. angusta*, hyacinths of the tallgrass prairies (see Chapter 10).

Cammassia cusickii is strongly erect and clump-forming under meadow conditions but, unlike *C. quamash*, it fails to set seed. To increase bulbs, lift and divide them before late-summer mowing and spread them around in autumn (planting depth 10cm/4in).

LILIES

I have not done well with *Lilium* under meadow conditions. As they remain green right through to late autumn, they do not fit into a grass-cutting regime. Neither do they enjoy the competition of a dense grass sward. I failed with *L. pyrenaicum*, which is such a success in the north of England and Scotland. *L. martagon*, to my pleased surprise, has seeded itself into a steep, grassy bank and has done well for several years. Being a willing colonizer in our climate, it should be an excellent candidate, though I have scarcely experimented as yet. It will be happiest in a semi-shaded position where the turf is thin. Typically, this turkscap has flowers of a murky purple. There are improved cultivars, as well as a scintillating albino, but these are hardly likely ever to be available in great enough numbers to be experimented with in turf.

The triplet lily, which we used to grow as *Brodiaea laxa*, is now *Triteleia laxa*. In its best colour forms it is deep blue, the funnel-shaped flowers borne in loose umbels about 30cm/1ft tall. Planted dormant in autumn, it takes a long time to reach its flowering, late June to July. In a border the lank leaves are a slight liability, but this is unnoticed in a meadow and it is doing well in one of our driest, poorest areas, in full sun, where blue is just what's needed at that season. But it is early days. We have not long been experimenting.

IRIS FAMILY

Some of the rhizomatous irises are well adapted to life in wet meadows. I have our native yellow *Iris pseudacorus*, of course, but round the fringes of ponds. (The North American equivalent is the blue flag, *I. virginica* var. *shrevei*, which grows in marshes in the tallgrass regions.) Near to the horse pond I have also established the bluish-purple *I. sibirica*, but I fear it is weakening.

Much more obviously suited to meadow conditions is the bulbous *Iris latifolia* (*I. xiphioides*), known as the English iris, though it originates from the Pyrenees. It has wide, flat blooms, purple in the best strains, though also pale mauve and grey. It is very showy and does not flower till the end of June. In borders, it increases prodigiously, so there is frequently spare stock to transfer to the meadows and at Dixter there are two areas where it is concentrated, the upper moat being one of them. The bulbs do not die out, but their freedom of flowering varies a good deal from year to year. They make a striking impression, as their flowers are large and a few go a long way. John Sales grows it, far more successfully than I, in his Gloucestershire garden.

Lilium pyrenaicum flowering in June in a meadow by the loch at Frenich in Perthshire, Scotland. This lily clumps up and is long-lived, especially in gardens cooler than mine in Sussex. Because so easy – I have seen it surviving in otherwise derelict gardens in the northern British Isles – it tends to be sneered at by *Lilium* aficionados. Its airborne scent is distinctly rank and unpleasant.

Brightest of spots in our late-May and early-June meadows are provided by the magenta *Gladiolus communis* subsp. *byzantinus* (1m/3¼ft). It is one of those corms that has been around in our borders for a great many years, increasing freely from cormlets, though just about sterile as to seed production. If you order this from a bulb firm, you are likely to be supplied with much paler-flowered, not nearly so exciting goods. Some small enterprises have the version we want. Beth Chatto offers it.

This gladiolus cannot increase in the meadows, except by clumping up a little, which it does, reliably maintaining itself. It is of a striking colour for which we are particularly grateful in this setting, acting as a balance to the masses of buttercups and contrasting satisfyingly with green grass. Spreading it around is a question of remembering to do the right thing at the right time: harvesting some corms in summer, when dormant, and planting them into the meadows as soon, after cutting, as the ground has become soft enough to make planting easy. Young leaves start to appear in the autumn, so this is not a gladiolus for harsh climates.

CYCLAMEN

It is only lately that I have tried the hardy, autumn-flowering *Cyclamen hederifolium* (8cm/3in) as a meadow ingredient, underneath a couple of *Crataegus* at the edge of the orchard, where the turf is thin, although it has to be shared with a vigorous colony of cow parsley (*Anthriscus sylvestris*). But the two of them have opposite growing seasons: the cow parsley's in spring, the cyclamen from autumn to early spring. It is the grass-cutting regime that presents the main problem. If the season is wet, the cyclamen starts to flower in late July – not till several weeks later, if dry, as it often is beneath trees. So the grass must be cut before this, which is before we are ready to cut the rest of the orchard. Still, we do it in that area. The grass needs a second cut in the autumn, by which time the cyclamen is in leaf and flower, so we have to mow around the plants as vigilantly as we can. They manage this all right at Kingston Maurward, near Dorchester in Dorset, and there are enormous old cyclamen corms in the grass in that area. What chance self-sown seedlings will have, I have yet to learn, though Beth Chatto tells me we must search for new-borns among the leaves in early spring, collect them and plant in trays. This she does in her grassy wood, which is mown close in late summer. 'After three years they are about the size of a hazelnut, large enough to pot individually,' she writes. 'When they have produced strong roots and leaves, we plant them out, adding to the drifts already established.'

The cyclamen corms need to be planted only a very little below the soil surface. The flowers are normally some shade of pinky-mauve, though we have pure white ones also and they breed true from seed. The leaves are a great asset, from autumn to spring, being beautifully marbled in shades of green.

Vivid magenta spikes of *Gladiolus communis* subsp. *byzantinus* look splendid in a setting of buttercups and cool grass stems. In a meadow community it holds its own, but seldom sets seed and cannot therefore spread itself. In a border stock builds up rapidly and can then be transferred to your meadow.

9. The Cornfield Scene

In a widely held misconception of what a meadow should look like, the beholder visualizes a brilliant haze of scarlet poppies in a sea of shimmering green. When such displays of gay flowers are seen in open country, there is such carefree abundance in their colours and extent that even the most hard-bitten road-hog may momentarily cast them a glance, although, of course, without slackening speed.

These images of breathtaking colour are translated in the beholders' minds to a similar, though in scale much reduced, replica in their own gardens. Which is perfectly feasible, but what they have seen is not a meadow at all; it is a cultivated area in which a crop has been or is being grown, but where lack of weed control has allowed a mass of uninvited interlopers to benefit from land treatment which admirably suits their requirements. The crop is usually corn, meaning varieties and species of *Triticum* (wheat), a grass of the *Poaceae* family whose seeds are ground for making flour, but may also be barley (*Hordeum*) or oats (*Avena*).

The interlopers are annual cornfield weeds, the 'idle weeds' that Shakespeare lists in *King Lear* as growing 'in our sustaining corn' – weeds, because they are competing with the desired crop. The word 'corn' commonly occurs in their popular names: cornflower, corn cockle, corn marigold and many more. An old-fashioned agricultural system such as

Cornfield 'weeds' – white *Anthemis arvensis*, blue *Centaurea cyanus*, yellow *Glebionis segetum* and scarlet *Papaver rhoeas* – are carefully managed to make a summer display at the Old Vicarage, East Ruston, Norfolk. Alan Gray and Graham Robeson exclude corn cockle, *Agrostemma githago*, partly because the plants are too dominant and partly because its early flowering season means that it looks horrid when the rest are at their peak.

still exists in the poorer countries of the European continent, and elsewhere in the world such as South Africa, supports a wonderful annual flora among its arable crops. This was common in Britain too, but with the intensification of agriculture midway through the twentieth century and the introduction of herbicides, the cornfield weeds quickly disappeared. Many are now hard to find, as are the insect and bird species that depend on them for food and shelter.

The most serious deception is in thinking that these 'weeds' are anything more than ephemeral. Indeed it is their transience that drew painters such as Monet, Van Gogh and Childe Hassam to capture impressions of their fleeting beauty. So you can't just sow them in your garden and expect a repeat performance year after year without further effort from yourself. You can certainly specialize in cornfield weeds without bothering about the corn, but annual input in the way of cultivations and, most probably, of seed sowing will be necessary to maintain a balance of species.

And you will not be creating a meadow. Terminologically this word has accumulated a variety of interpretations, but in all there is the presumption of permanence and stability; of an area where there is turf and this turf is only ever superficially disturbed. Whereas disturbance in the form of cultivation (or climatic disturbance as with seasonal drought) is the essence of a cornfield and of the flora associated with it. However, because no other description sounds well on the ear, I shall contradict myself and refer to annual meadows. They are attractive and have validity in their own right, so we need to understand how this is achieved, how maintained and what the contents are likely to be.

THE REWARDS OF EPHEMERAL PLANTINGS

Many annuals have a quality of colour and freshness that is especially gladdening. 'Here we are,' they seem to say; 'enjoy us while you may. We cannot stay for long.' In the context of a meadow, such fleeting effects can be thrilling, if you are prepared to make the effort, every year, of cultivating, raking and weeding.

Annuals do not need the poor soil that is high on the priorities of the species-rich permanent meadow. On the contrary, they benefit from fertility and do best in moisture-retentive soil. In the purlieus of a garden, where there have been previous cultivations or where fowls or other domestic animals have been run, fertility can generally be assumed. Fertilizers, should they be needed, must be strong in phosphorus and potassium and low in nitrogen.

Another point in favour of annual meadows is their flexibility and spontaneity. The results are dramatically fast – there is no waiting for several years as with a perennial meadow – and rich colour appears within months of sowing. It is quite widely supposed that annual meadow ingredients can be included in the initial seed mix when creating a permanent meadow so as to provide colour in the first season, these to give way to the perennials in subsequent years. As I have said, it is better, I think, to be strong-minded and not to include annuals, whose shading of the nascent perennials in their first season of establishment may not actually kill but will certainly weaken them.

So, if you are creating an annual meadow, be prepared to go for it every year. You can experiment with seed mixtures, ringing the changes from one year to the next, and sweeping away any less-than-successful effects along with the remains. With annuals you could say that it is a crime to grow the same kinds year after year.

GETTING AN ANNUAL MEADOW GOING

The site *must be wide open*, the ground well prepared to a depth of 15cm/6in and weed-free. Sowing can be in either autumn or spring, though spring seems to be the better choice, as it gives greater opportunities for weed control beforehand. Weeds are the bane of annual meadows. Perennial weeds compete on a permanent basis, while annual weeds germinate and grow under the ideal conditions provided, and at the same rate as, or even faster than, the annuals we have sown ourselves.

Autumn-sown annuals will make strong plants, which will start flowering in early summer (May), but may let you down at the end of the season. Spring sowings (generally there is no advantage in sowing before mid-April, although it can be a month earlier on light soil in the south of England) will start flowering in June.

Seed is expensive and the urge to sow too thickly must be resisted. If you need an immediate knock-out impact as you approach your meadow, you may want to sow more thickly near to its margins, and denser sowing would be understandable if a plot were quite small. But over larger areas seed can be sown at 3g per square metre/ oz per square yard.

It is essential to sow evenly. This will be helped by bulking out the seed with a neutral carrier such as dry sand or barley meal, in the ratio of 10 per cent seed to 90 per cent carrier. The seed can be mechanically broadcast by a hand-held distributor; or, on a larger scale, by a tractor-mounted spreader.

Sowing can be followed by a light raking or harrowing, though this is generally unnecessary and its omission obviates the danger of burying the seed too deeply. Rolling is generally unnecessary, too, except on very light soils, liable to dry out.

At some point in the cycle, dead remains will need to be removed. Some people prefer to do this in the autumn, ploughing and cultivating and then leaving the ground open during the winter, towards the end of which a herbicide can be applied. This will catch perennial weeds like the low, dense, mat-forming creeping fescue, *Agrostis stolonifera*, a grass which establishes itself at the base of the crop in summer, but is exposed when the latter is removed. After a two-week interval, it will be safe to sow. Others prefer to leave the dried plant remains through the winter, claiming that they have a beauty of their own and look better than bare ground. That may or may not be true! In an urban setting they will collect all sorts of debris.

CORNFIELD WEEDS

A typical cornfield mixture comprises five species. Only five, you query? But many of a few is often more effective (and gives each component a greater chance to succeed) than a few of many. A standard mix usually consists of corn cockle, corn chamomile, cornflower, corn marigold and common or field poppy. Miriam Rothschild's 'Farmer's Nightmare' mix comprises all but corn chamomile among barley and oats. If you have a particular favourite of your own, you can add it. However, ten different species is almost certainly enough to be sowing even on a field scale, while in a small plot three will be far more effective and successful than a complex mix, in which a number of species are sure to be muscled out and hence wasted. In choosing your three, however, you must give thought to a spread of seasonal interest and to the length of flowering season.

Agrostemma githago, the corn cockle, with bright magenta-pink flowers, is tall (up to 1m/3¼ft) and apt to be dominant. It is included in all mixes offered, but if you make or specify your own, it may be better to reduce the amount of seed or even leave it out. Another disadvantage is that its season is short and that it spoils the meadow's overall appearance in later summer. (The same may be said of the tall lacy scorpion weed, *Phacelia tanacetifolia*, which has finely divided foliage and makes a haze of blue but has a short, early season and is an eyesore thereafter.)

Of cornflower or bachelor's button, *Centaurea cyanus* (20–50cm/8–20in), it is desirable to grow a wild strain, as most cultivated strains, which come in a wide range of colours, are too large and bulky and look wrong in a cornfield native-only setting; it must in fairness be added that the competitive conditions will much

A planting of cornfield 'weeds', cornflowers (*Centaurea cyanus*) and corn marigold (*Glebionis segetum*), creates a rich tapestry of flowers in this annual meadow.

Attractive weeds take over a field in Hungary where the crop has been taken: the beautifully constructed white umbellifer *Orlaya grandiflora*, spikes of viper's bugloss, *Echium vulgare*, and red field poppy.

reduce the vigour that you would find in a garden border. The season of cornflowers is surprisingly long.

Individually the plants of common or field poppy, *Papaver rhoeas* (60–75cm/2–2½ft), which is another European native that has naturalized throughout North America, are not very long-lived, but their seeds germinate over a period, so that there is a succession of bloom. The translucent papery petals frame a ring of almost equally beautiful stamens surrounding a central stigma. Additionally, their seed heads are attractive, 'supremely graceful urns', as Celia Thaxter mused in her island garden.

Corn marigold, formerly *Chrysanthemum segetum* (30–60cm/1–2ft), now correctly *Glebionis segetum*, has brilliant yellow daisies, which are scattered as in a constellation, and rather glaucous foliage. It makes a large plant (so less seed is required), has a very long season and the seed is easy to collect. A winner, in my opinion. I have no personal experience of the behaviour of corn chamomile, *Anthemis arvensis*, a small, white, yellow-centred daisy, but it certainly contributes nicely to a tapestry effect.

One of the most effective deliberate cornfield displays that I have seen is at the Old Vicarage, East Ruston (next to the church), near Cromer, in East Norfolk. Here (beyond the confines of an elaborate garden), the owners, Graham Robeson and Alan Gray, omit corn cockle from the standard mix. In fact, they adjust their mix, varying the proportions of each ingredient according to the performance of the previous display. If, for instance, cornflowers were weak, they add more of that. In some cases they need no seed at all, as there is a sufficient seed bank in the soil from previous seasons. In annual meadow displays in public parks and gardens, however, such a risk is rarely taken and sowings are repeated annually, willy-nilly.

At East Ruston, the crop is cut and carried in October, after flowering, and a month later the area is treated with glyphosate to deal with perennial weeds (mostly grasses). The land is then ploughed and harrowed. Spring cultivations are generally unnecessary unless the ground has been seriously panned over winter. Light cultivation in spring, to break up the pan, will then be necessary. That applies to all cornfield cultures where annuals are dominant.

WEEDS, WEEDS, WEEDS

To harp on the subject of competing weeds may be off-putting to gardeners who have a romantic vision. I don't want to put you off but, all the same, weed control, operated from the start, is a primary requirement for success. Weeds can be divided into two kinds, perennial and annual.

Allowing the land a fallow period, during the summer growing season, is ideal from the point of view of controlling perennial weeds like docks and nettles, bindweed, couch grass and thistles, before the enterprise is entered upon. When they are growing strongly, spray them with the translocated herbicide glyphosate. More than one application will be needed and can be applied when the weeds that have escaped are growing again. This herbicide is most effective when the weeds are growing strongly, in the summer, but can also be used at any time when they are green. Therefore, if a spring sowing of meadow annuals is preceded by autumn cultivations, subsequent weed growth can be treated with glyphosate up to late winter (February in the UK). This also applies to an established annual meadow regime and should, as already emphasized, always precede spring sowings.

Cultivations actually promote the spread of certain perennial weeds, broken pieces of whose roots are thereby distributed and make new plants. Couch grass and creeping field thistle are particular cases in point. Couch can probably be caught by the autumn and late winter herbicide applications. The thistle comes late into growth and may need spot treatment, in early summer: wipe the shoots with a glyphosate-impregnated glove. Annual weed seeds may be present in large quantities in the topsoil. Allowing them to germinate while the land is fallow and treating them with herbicide will not deal with the many seeds which have remained dormant, nor with those which are turned up to the surface in the course of cultivations.

One of the most successful methods devised for coping with this situation has been investigated by Dr Nigel Dunnett at the Landscape Department, University of Sheffield, and I am indebted to his work for much of what follows. If a mulch, in this case a layer of sand, is spread to a depth of 1cm/½in over the prepared seed bed prior to sowing on top of it, without further cultivations, germination of weed seeds is reduced to negligible proportions.

If we now suppose that you did not provide this treatment and that a daunting crop of groundsel has come up alongside your own annuals, overtopping them, the groundsel itself can be topped at, say, 10cm/4in above soil level, while leaving your lower-growing crop untouched. This can be done with a strimmer, choosing the right moment when the groundsel is in bud but not yet flowering. Where all-enveloping chickweed is the problem, I think we must just hope that our annuals can grow alongside it without too much detriment to their performance.

NATIVE OR NOT?

Purists who insist that gardeners must use locally specific plants to preserve genetic integrity are especially vociferous when it comes to annuals. Dunnett suggests that alien annuals are acceptable for distribution in built-up or confined areas, such as enclosed gardens or the public green spaces in a city like Sheffield, where he operates. 'Why', he argues, 'does a non-native species that we are happy to see growing in a flowerbed suddenly become a menace in a meadow setting?' A cautionary note needs to be sounded here: the danger of introducing an attractive annual which will find conditions so much to its liking that it will spread beyond the confines of your plot and become a menace to the countryside as a whole. In every country you can visit, examples of this kind are easily spotted.

Dunnett's arguments are understandable, in the circumstances, and designed as a riposte to the holier-than-thou ecologists, who insist on an untainted native flora. My own feeling is that, with modern communications being what they are, unfortunate escapes are inevitable. There may be something we can do about stemming individual cases or threats in their early stages, but generally we are alerted too late and might as well admit as much. To lay down a blanket policy is not practical, although it may be political. If growing species native to your locality sounds right to you, go ahead, but it is well documented that plant communities migrate according to climatic conditions, and the seed is carried by wind, birds and animals, even by car tyres, beyond its normal boundaries. Indeed, many annuals native to Europe have naturalized throughout the USA. The common poppy is but one of them.

OTHER ANNUALS TO GROW

If you do not just stick to native seed mixes, what other choices are open to you? If you make your own mix, you can ensure that it includes both long-flowering species (like corn marigold) and late-flowering kinds (like the coneflower, *Rudbeckia hirta*, and a mix of tall *Cosmos bipinnatus*), so the season can be extended into autumn. You can also vary, by selection, colour combinations as well as the height of the contents. It is good to include some annuals (not too many) that rise head and shoulders above the generality, such as orach and sunflowers. In the USA mixes are formulated for specific regions and these will help you devise your own. It is worth repeating that a few species will be far more effective and successful than a complex mix, especially on a small scale.

LONG-FLOWERERS

Ammi majus is a lacy umbellifer native to north-eastern Africa and Eurasia with purest white flowers. (It is similar in appearance to *Daucus carota*, the wild carrot, which is considered a pest in much of the USA.) Height is variable: from a spring sowing not more than 1m/3¼ft, but twice that height from an autumn sowing

FAR LEFT Blue-flowered borage, *Borago officinalis*, self-sows enthusiastically and has quite a long season. It combines well sown with yellow *Rudbeckia hirta*.
LEFT Orange pot marigold, *Calendula officinalis*, is a hardy annual with a pleasing aroma. The rays are excellent – and colourful – in a salad.
BELOW In your own garden, it is easy to get a tapestry effect like this with a spring sowing of the red form of annual flax, *Linum grandiflorum* 'Rubrum', and the tall purple candelabra of *Verbena bonariensis*.

Ammi majus is one of the purest of white cornfield weeds, with well-spaced umbels that give it a free-and-easy look. As well as being a good foil to bright poppies and cornflowers, it is effective with quieter colours. Beloved of flower arrangers, it is seen here in Sarah Raven's Sussex garden.

grown in a border. This is a Dunnett favourite and one of mine also. It has quite a prolonged summer-flowering season and makes an effective foil for the bright colours of other annuals. Seed needs to be fresh, as with many other umbellifers, notably with *Orlaya grandiflora*. White and as beautiful as the last with the added attraction of flowers with petals of unequal length, the longest pointing outwards, this made a great spectacle in some fallow Hungarian corn meadows in June, especially when mixed with scarlet poppies and blue viper's bugloss, *Echium vulgare*. Once you have fresh seed you must save your own and work up stock.

To grow alongside blue cornflower, Dunnett speaks highly of Mexican poppy, *Argemone mexicana* (60cm/2ft) and red flax, *Linum grandiflorum* 'Rubrum' (60cm/2ft), native to north Africa and southern Europe, but naturalized throughout the USA. The former has white-veined, thistle-like leaves and yellow or orange flowers from midsummer, followed by fruits. I use the red flax for bedding out and like it immensely – deep-red flowers with a dark rim and faint radiating striations from the centre. It has a very long season.

Another long-flowerer of the same height but a far bulkier habit is the California poppy, *Eschscholzia californica*, which is, typically, dazzling orange. Other colour strains have been developed, gradually reverting, as they self-sow, to orange. Let it be orange, I say. Try it with red flax and *Ammi majus*.

Besides *Glebionis segetum*, the corn marigold, already cited, forms of other similar annual chrysanths (native to Mediterranean regions) are suitable: *Chrysanthemum carinatum* (now *Ismelia carinata*) and *C. coronarium* (now *Glebionis coronaria*). Dunnett combines a cultivar of the latter, 'Primrose Gem', with *Nigella damascena* 'Oxford Blue', a dark blue love-in-a-mist (75cm/2½ft).

Blue-flowered borage, *Borago officinalis*, has quite a long season and contrasts well with orange pot marigolds, *Calendula officinalis* (both 30–60cm/1–2ft). When they self-sow, the bright orange, daisy-like flowers become progressively less double, but no matter in this context.

QUICK-GROWING, EARLY-FLOWERING FLIMSIES

Species with a thin texture, whose presence will present no problems to later-flowering mainstays, should include the jolly little toadflax or baby snapdragon, *Linaria maroccana*, native to Morocco, which has naturalized throughout Europe and the USA. Its tiny flowers along upright stems come in a range of colours. Annual baby's breath, *Gypsophila elegans* (30cm/1ft), a native of eastern Europe that has naturalized throughout the eastern USA, casts a white cloud within five weeks of sprouting and can be regarded in the same light. As can viscarias in blue and pink (really *Silene coeli-rosa*) and California bluebell, *Phacelia campanularia*, which flowers for three weeks and fizzles out by early summer.

Baby blue eyes, *Nemophila menziesii*, native to California, is a short (25cm/10in) early-season annual with sky-blue flowers that finish blooming before midsummer. The sweet William catchfly, *Silene armeria* (60cm/2ft), native to Europe, lives up to its American common name of none-so-pretty, with masses of rose-purple flowers in clusters. Small-flowered catchfly, *S. gallica* (30cm/1ft), known as the windmill catchfly in the USA, resembles a small white campion, the flowers borne singly and up to 1.5cm/½in across.

Besides *Papaver rhoeas*, already cited as standard in any cornfield mixture, there are the cultivated hybrids, Shirley poppies. There is also the long-headed poppy, *P. dubium*, which takes the place of the field poppy in northern and wetter areas of the British Isles, but is a rather less intense shade of red. Other poppies worth considering are opium poppy, *P. somniferum*, and ladybird poppy, *P. commutatum* (45cm/1½ft), with deep crimson flowers and a black spot at the base of each petal. Dunnett recommends combining the latter with purple orach and purple millet, *Panicum miliaceum* 'Violaceum'.

The hollow-leaf annual lupin, *Lupinus succulentus* (45cm/1½ft), offers early colour – mostly blue – in North American meadows but forget Texas bluebonnets (*Lupinus subcarnosus*) unless you live in Texas. The fields of bluebonnets made famous by Lady Bird Johnson, which inspired her to form the National Wildflower Research Center at Austin (now the Lady Bird Johnson Wildflower Center), depend on a long season and extreme heat in June. Away from home they are slow to perform.

LATE-FLOWERING ANNUALS

Among the best of late-flowering annuals is the American *Coreopsis tinctoria* (1m/3¼ft), yellow with a bronze zone, or red on yellow. Dunnett mixes this with *Verbena bonariensis*, native to South American prairies, which I cherish as a short-lived perennial gate-crasher in the borders at Dixter, to the amusement of friends from New South Wales for whom it is a noxious weed. Of similar height to the coreopsis but with a longer season is our old friend *Cosmos bipinnatus*, in mixed pink, crimson and white colours. And a third American daisy flower, with a long, late season, is the deep yellow coneflower, *Rudbeckia hirta*. The only snag with this, among taller partners, is too short a stature (70cm/2 ft) for satisfactory visibility. But taller strains are available. Dunnett recommends mixing the coneflower with coreopsis and corn marigold for a long-lasting show of gold. The Tahoka daisy, *Machaeranthera tanacetifolia*, is yet another American, even shorter at 45cm/1½ft and with aster-like mauve flowers with yellow centres.

Dunnett is keen on the shoo-fly plant, *Nicandra physalodes*, and so are many of my friends. It is bulky, 1m/3¼ft tall, weed-suppressing and has a long season, its ripe seeds remaining as pleasing skeletons into winter. It is a curious-looking plant which I find faintly disagreeable, but that is my problem.

Indian blanket or firewheel, *Gaillardia pulchella*, native to hot, dry regions of

North America, with bronze daisies, the rays yellow-tipped, is a stemmy annual (60cm/2ft). It is disappointing in a border, I always think, but ideal in our context and easily established from seed.

Many annual grasses add a delightful wispy quality to a meadow and are good late on. They often dry well on the plant. There is the foxtail grass, *Setaria macrochaeta*), with nodding bottlebrush flower heads, bleaching in autumn. Purple millet, *Panicum miliaceum* 'Violaceum' (1m/3¼ft), is topped in late summer by arching, purple flower heads and remains in fine shape for much of the winter. Annual quaking grass, *Briza maxima* (60cm/2ft), will reappear year after year from self-sown seedlings; indeed it can become a nuisance.

TALL ANNUALS

For a tall annual with a long season, you cannot beat purple orach, a reddish-purple strain of *Atriplex hortensis* (related to spinach). It will reach 2m/7ft on good soil and remain in excellent condition to the end of August. Even in seed it is still presentable. And it is a great self-sower, from year to year.

Sunflowers, *Helianthus annuus*, are an attractive concept. However, those grown commercially, carrying one flower on each plant, have far too short a season for our purpose. What's needed is a smaller-flowered cultivar with a branching habit. We find that 'Valentine' (1.5m/5ft), with pale yellow flowers and a dark disc, has great staying power. Dunnett recommends cucumber-leaf sunflower, *H. debilis*, and its cultivar 'Vanilla Ice', both of which are compact, multi-stemmed plants.

Two larkspur species that are native to Europe and common around cornfields in Hungary, both of them purplish, are *Consolida regalis* and *C. orientalis*, the latter spectacular. Both 1m/3¼ft tall. *Cosmos bipinnatus*, already cited, is a native of Mexico that can easily adapt to all regions of Europe and the USA. Rich soils tend to produce lanky plants, which is fine in our context. At a similar height, 1.2m/4ft, the rose mallow, *Lavatera trimestris*, native to the Mediterranean region, blooms at the same time as the cosmos and in the same colours – white and pink – should such polite colour theming be to your liking.

Eastern larkspur, *Consolida orientalis*, making a startling display next to a wheat field in Hungary. Where a selective herbicide has been used the field is 'weed' free, but the larkspur dominates in an adjacent area where the ground has not been sprayed. By sowing annuals in contrasting and complementary swathes and strips, similar effects can be achieved in gardens.

10. Prairies

North American prairie plants are of the greatest importance to gardeners. Even though the climate that gives rise to prairie vegetation may be very different, many of our most widely grown species, or derivatives from them, are mainstays of our borders. Adaptation to low rainfall and low winter temperatures makes them hardy in most areas. Generally they need the deep, moist, slightly alkaline soil found in their own regions, and an open sunny situation.

Prairie, meaning meadow, was the name French explorers used for the vast treeless landscape they found stretching through middle America. It describes grasslands of a fairly permanent nature (in contrast to meadows, which are always in a state of transition), their sustainability being on account of either evapo-transpiration exceeding rainfall – conditions which do not allow trees to grow, except where they find the extra moisture of river beds – or regular burning in spring. (Equivalent areas in other parts of the world are known, variously, as llanos, pampas or steppes.) Grasses with deep questing roots, however, can survive the hot sun and evaporating winds and perhaps 80 per cent of prairie consists of grasses, mostly of a clump-forming rather than a mat-forming habit. In between grow the forbs.

The forbs – the herbaceous perennials, biennials and annuals – include many genera,

The strong-growing, boldly upright North American perennial culver's root, *Veronicastrum virginicum*, is a notable success in the Dixter prairie, reaching 1m/3ft. This is the white-flowered form, seen here in late summer among bleaching grasses. I collected the seed from the Black Dog Reserve in Minnesota some years ago, and it has established well in fertile soil.

such as *Echinacea, Helenium, Monarda, Rudbeckia* and *Solidago*, that have long been a staple diet in our gardens. They provide colour in late summer and interest for much of the winter, the skeletons of seared plants remaining in beauty over many months.

NORTH AMERICAN PRAIRIES

About one fifth of the North American land mass, most of it lying between Texas in the south and Manitoba in the north – the great plains west of the Mississippi river and east of the Rockies – was occupied by prairie, or a matrix of prairie and woodland. This is now reduced to 1 per cent; some say, to one tenth of 1 per cent.

Seeing that such large areas are involved, it is natural that both grasses and forbs vary greatly over different areas. These have, for convenience, as well as for natural reasons, been divided into four main sectors. An imaginary north-to-south line is drawn along the 100th meridian (line of longitude). Another is drawn east to west along the 39th parallel (line of latitude). This divides the entire plain area into four quadrants. The western quadrants become increasingly dry as they approach the rain shadow of the Rockies. There they are shortgrass prairie, say up to knee height, though they become longgrass prairie where there is extra moisture in swales (depressions). Tallgrass prairie dominates in the eastern quadrants.

Besides high evapo-transpiration, regular periods of drought and grazing (by bison and elk, for example), fire has been a factor in sustainability, since it disposed of most woody seedlings, so that grass could continue

Dry, scrubby grassland in California with native xerophytic plants, including spikes of Our Lord's candle, *Yucca whipplei* subsp. *parishii*, and trails of hair-like dodder, *Cuscuta*, which is wholly parasitic, without roots, and gains sustenance from plants around it.

to dominate. Before man's arrival, it was caused by lightning. Thereafter, it was often deliberate, having a number of beneficial results. Fires are still necessary for the preservation of most types of natural prairie. Inevitably they destroy wildlife and must first be carefully timed, to do the least undesirable damage, and second be limited in extent, perhaps burning only one quarter of any one remnant prairie at a time.

Some trees, especially a range of oak species, are notably fire-resistant. Where, thanks to their tough bark, they survive, notwithstanding fire around them, savanna develops, this being grassland dotted with a limited selection of trees. The savanna is a transition zone between prairie and woodland, and is particularly common on the eastern prairie/forest border. Many of these oaks also have the interesting self-preserving capacity for developing a scrub of suckers from their roots. This scrub forms a wonderful habitat for migrant birds and for much other wildlife. Where fire ceases – often a large river forms the divide – forest takes over. There are forest trees on the margins of all the prairie areas of North America.

Prairies have been classified into three categories – wet (hydric), moist (mesic) and dry (xeric) – by available soil moisture. Mesic is intermediate between wet and dry. It occurs on sites that are relatively well drained, but have high moisture available through most of the growing season, and includes the most fertile soils. As it is the easiest to convert to crop production, it was the first type of prairie to disappear and only small remnants remain. A few of the hundreds of wildflowers include big blue-stem (*Andropogon gerardii*), lead plant or prairie shoestring (*Amorpha canescens*), prairie blazing star (*Liatris pycnostachya*), showy goldenrod (*Solidago speciosa*) and switch grass (*Panicum virgatum*).

Wet prairie, when drained, became equally productive, so that went next. Typical wet prairie plants include iris (*I. virginica*), false aster (*Boltonia asteroides*), prairie cord grass (*Spartina pectinata*) and various sedges.

Only dry prairie on slopes and well-drained uplands continued to be sufficiently intractable partially to escape man's attention and conversion to purposes entirely different from native prairie. Dry prairies that remain tend to have an abundance of little bluestem (*Schizachyrium scoparium*), blue sage (*Salvia azurea*), pale coneflower (*Echinacea pallida*) and prairie or Carolina larkspur (*Delphinium carolinianum*). They were often grazed, and all types of prairie were usually cut twice a year for hay.

PRAIRIE RESTORATION

Corridors of unspoilt prairie flora and fauna can sometimes survive along rivers and roadways (see Chapter 11). The widespread loss of prairie to development and agriculture has stimulated the restoration of these habitats, Jens Jensen being one of the earliest garden designers to advocate them. The Curtis Prairie at Madison, Wisconsin, was the first of the tall-grass prairies to be re-established, using remnants along the Wisconsin river as a source of native seeds and sods. Little of the original habitat was left when the fields were purchased in 1933, the 25-hectare/62-acre area having been converted to agriculture – corn, oats and pasture in rotation – in the mid-nineteenth century. The restoration goal was 'to replace pasture and ruderal [weedy] vegetation with species of native plants for use in research and education.'

It took more than thirty years for the Curtis site to begin recovering prairie soil characteristics, including the organic carbon lost to oxidization during the years of agricultural use. Mound-building ants, the major soil cultivators of tallgrass prairie, are present only in an unploughed section. Native fauna includes deer and coyote, but the site is too small to support the bison and elk that roamed the original prairies. (Grazing by ungulates alters the landscape by removing plant material, concentrating nutrients, trampling, and influencing decomposition.) Studies of the effects of fire treatments on selected plots gradually reduced the incidence of bluegrass (*Poa* spp., from the former pasture) while native prairie perennials and weedy forbs increased. Fire continues to be used – on average two out of three years – to control invasive species (such as biennial sweet clover, *Melilotus alba*) and curtail the encroachment of woody species. The use of herbicide is also being assessed. The Curtis Prairie is an on-going project, involving the assessment of plant composition, germination studies and the management of grassland in an area of increasing urbanization. Many factors continue to challenge the restoration ecologists.

Beth Chatto and I, guided by the extremely knowledgeable Colston Burrell, visited the restored prairie at the Minnesota Landscape Arboretum in a wet area south of Minneapolis called Black Dog Reserve. It lies between the confluence of the Minnesota and Mississippi rivers, and is strewn with boulders – reasons for its never having been converted to farmland, as nearly all prairie has. This was on 17 October, and the breath of winter was already evident, with frost at night, but fine weather and warm sunshine. Of the many grasses, the most familiar was the hazy *Panicum virgatum*, which comes fully into its own in autumn. Some of the plants we walked through were way above our heads, all dry, but making beautiful skeletons. There was no trace, now, of the low-growing, spring-flowering forbs, such as a deciduous violet. Visible remains of taller species included, among much else, *Helianthus* (including what we call Jerusalem artichoke, *H. tuberosus*); *Rudbeckia* (including *R. laciniata*); *Silphium* (of which *S. perfoliatum* is the most familiar in our gardens, though *S. laciniatum* and *S. terabinthinaceum* have claims to be superior); *Monarda* (bergamot), including *M. fistulosa; Veronicastrum virginicum* (of the characteristically whorled leaves); and *Verbena hastata*, of which I grow the pink-flowered cultivar 'Rosea'. It was a kind of home from home, and, it is worth adding, once you have seen a familiar species in its native habitat, it comes fully into focus and you never forget its provenance.

ESTABLISHING A PRAIRIE

Like creating a garden of perennials, prairie plantings can take a few years to become established, requiring diligent maintenance, but thereafter become more self-sufficient. In residential areas, to be successful, a prairie

The small-flowered sawtooth sunflower, *Helianthus grosseserratus*, has established well in the Dixter prairie. It spreads vegetatively by means of a running rootstock and can compete with our natives, like the lesser knapweed, *Centaurea nigra*, which still has masses of mauve-purple flowers in September.

should be planned with a cohesive layout that takes into account existing features, such as the house, shed, boundaries and the lie of the land. In large, rural situations, leave space for a firebreak. As described in Chapter 3, mown grass and edgings of brick or stone indicate that an area of grasses and wildflowers is intentional, albeit non-traditional.

There is no right way to get started. Colston Burrell, who restored the prairie at the Minnesota Landscape Arboretum, and Neil Dibbol, who runs a prairie nursery in Wisconsin, consider a blank canvas to be critical to success. All turf and weeds (in the USA these include cool-season grasses, such as quack, *Elymus* (syn. *Agropyron*) *repens*; bluegrass, *Poa*; and brome, *Bromus* spp.) must be removed, they say, either manually or with herbicide, in order to eliminate competition for water and nutrients. What Burrell calls 'sod stripping' is practical only for small areas and has the disadvantage of removing topsoil along with the turf roots. Smothering grass and weeds with black plastic takes a full growing season to be effective. Herbicide, then, in a light application is the easiest way to kill most lawn grasses. Stubborn perennials weeds may require several applications or removal by hand.

Others experts advise plugging young transplants into what is already growing, having dug out the undesirables.

PLANTING THE PRAIRIE

On a large scale (0.2 hectare/½ acre or more), seeding is the least expensive option and my advice is to prepare the site according to the instructions provided by your supplier. Whatever you do, don't try a mix of seeds and transplants since seeded areas must be mown regularly for the first two years whereas transplants do not take to being mown down except when they are dormant in spring. Because most of my readers will have smaller spaces available, I shall concentrate here on using seedlings and plugs, and start, as before, with my own, albeit limited, experience.

I brought home a lot of seed from Minnesota, with the notion of trying to create my own little prairie in a piece of meadow (once an orchard) in my Sussex garden. Determined to give the seedlings a good start, I raised them in containers, then lined them out to grow on in cultivated ground. Only when they had made strong plants were they transferred into pockets of prepared ground in the designated prairie. Conditions, here, were highly competitive. British winters are mild, snow cover if any being of short duration. The meadow into which I was planting comprises a dense sward, green and growing year-round with many creeping grasses and forbs. It reaches a flowering peak in early summer. This was largely running to seed, looking sad and past it, by the time the prairie flora was peaking. This is visually unacceptable and is the main drawback to prairie material being planted into a European-style meadow.

Prairie flora is totally dormant in winter. When it comes to life, in late spring, it is generally clump-forming (including the grasses) and most of it is unable to compete with the conditions in my garden. Some of my transplants survived for a few years but many vanished sooner or later. This included a range of solidagos,

Liatris spp. (which seem to make short-lived plants anyway – a pity as they are notably popular with butterflies), *Ratibida* and many more, which I prefer to forget, as the toll is depressing. Holding their own are, outstandingly, *Veronicastrum virginicum* and *Eryngium yuccifolium*, though they do not go so far as to self-sow. *Helianthus grosseserratus*, a fairly inconspicuous sunflower, has a running, tuberous rootstock and that is definitely spreading. None of the prairie grasses could cope with our native ones, which grow year-round.

My experience has led me to conclude that the way out of it has to be by creating a prairie pure and simple, eliminating the native competitors, as Dibbol and Burrell suggest. However, we have not given up by any means and, not being puristic, we include likely contestants from other parts of the world: pampas grass, *Cortaderia selloana* 'Pumila', for instance, and *Galega officinalis* 'Alba'. Full sunshine and exposure gives all these plants their best opportunities.

Still on the subject of prairies in the British Isles, where competition from native meadow ingredients is intense, I now want to move on to Dr James Hitchmough's particular study of prairie-style planting as it can be practised here, particularly in the south of England, where summers are reasonably warm. His principal aim – to learn how to create and manage very attractive tall herbaceous vegetation at low levels of capital and maintenance costs – has pertinence in other regions of Europe where winters are cold and summers hot. Such prairie plantings are visually attractive in late summer, autumn and for quite a slice of winter.

Hitchmough laid out his experimental plots on the site of some derelict allotments. Here he set out (1) to investigate the effect of sowing rate on the establishment of fifteen species of prairie perennials (other than grasses) and of two species of prairie grass; (2) to investigate the effect of soil surface type on the germination and establishment of the above, following a late-summer and an early-winter sowing; and (3) to investigate the influence of longer-term management (spring mowing, burning and overspraying with a contact herbicide) on the development of prairie plant communities.

The site was both weedy and full of weed seeds, so glyphosate was used to eliminate all that was currently green. There were three soil treatments: the native topsoil of the site by way of control; topsoil inverted to expose subsoil and topsoil with a 3–5cm/1½–2in much layer of coarse sand. It was a grit sand (such as is used for concrete making) and it locks together, so there is no blowing. All was scientifically randomized but I shall only give the gist of the experiment here, not the details. There were three sowing rates. Sowing took place in August and in early December.

On average, winter sowing gave better results than summer. The seedlings entering winter from the summer sowing were very small and many were destroyed in winter – washed away by rain or killed by soil heave due to frost. The winter-sown seeds do not germinate till spring.

There were fewest weed seedlings on the plots treated with a sand mulch, while, by comparison, the prairie seedlings did well, so that treatment is strongly recommended. In sum, winter sowing with a sand mulch gave best results.

TOP, LEFT TO RIGHT Tall tickseed (*Coreopsis tripteris*) and rosinweed (*Silphium integrifolium*), the last with oval leaves that feel like sandpaper, are among the plants that have proved to be less palatable to British slugs than many North American natives.
BOTTOM A lively late-summer display of purple coneflowers, pink beebalm (*Monarda fistulosa*) and yellow goldenrods (*Solidago ohioensis, S. rigida* and *S. speciosa*) among grasses in James Hitchmough's experimental plot at Sheffield University. *Echinacea purpurea*, with very purple rays, is the hardiest and easiest coneflower under British conditions. Many variants have been bred, including white-flowered kinds. Since they are nearly always raised from seed, there is a good deal of variation in colour and habit.

Examples of the species sown by Hitchmough on his trial plots include such familiar genera (sometimes species also) as *Baptisia australis*, *Coreopsis lanceolata* and *C. tripteris*, *Echinacea pallida*, *Liatris pycnostachya*, *Monarda fistulosa*, *Rudbeckia subtomentosa*, *Solidago ohioensis*, *S. rigida* and *S. speciosa*.

Slugs are a great problem, scarcely encountered by prairie plants in their native habitats, where winters are too cold for slugs to exist. They can be controlled by baiting with metaldehyde. This remains effective for about three weeks and must be used at the time when germination is just taking place, which would be in February following a winter sowing. This method of slug control is unacceptable to those averse to using chemicals. So Hitchmough conducted other experiments in an effort to find out which species may be grown without the use of slug pellets. He made a trial of seedlings of prairie flora and tested their palatability to slugs (he also made nightly slug counts, to learn more about their habits). *Silphium* spp. turned out to be wholly unpalatable, whereas others, like *Asclepias* and *Ratibida pinnata*, were disastrously susceptible.

MAINTENANCE

Because prairie plants do not start growing till late spring, when the soil is warming up, you can cut back the weeds in seeded prairies by mowing (with the blade set high, at 12cm/5in) for the first year or two without harming the wildflowers and grasses. In the third year and after, the grasses and forbs will begin to shade or out-compete the weeds, reducing maintenance. Even so, weeding – by hand or the careful application of herbicide – will be the most arduous job during the first summer and will continue to be necessary. Other than adding more plants, avoid disturbing the soil as this will produce a new crop of annual weeds.

Transplanted prairies are mown in early spring to make way for new growth, having been left over winter to provide interest and habitat for wildlife. This can be done as late as April, without endangering the prairie element. Raking off debris is essential. Alternatives to mowing include burning, for which you need a permit, or burning off with paraquat. To protect wildlife, burn less than half the prairie in one year.

Your cutting regime will also be influenced by conditions. The restored prairie at the New York Botanic Garden in the Bronx covers only 0.1 hectare/¼ acre and is wet, particularly in spring, so it is cut in November after seeds have ripened and dispersed, the debris being allowed to decompose on site. It is mown twice, in late April and early May, before the Turk's cap lilies and baptisias emerge.

PRAIRIE PLANTS

Your type of soil and its depth and drainage will have a strong influence on the sort of species you should be attempting. Then, to function as an ecosystem – a point Burrell emphasizes – you need to mimic the diversity of species found in native prairies. 'A prairie,' he writes, 'may have as many as 300 species but is usually dominated by three to four grasses and about 20 wildflowers. To approximate this diversity, your plant list or seed mix should have at least 25 reliable species of wildflowers and grasses.' To keep the scene lively, bear in mind flowering sequences.

Many prairie species have a very extensive habitat range and it is hardly surprising that they themselves, without being divisible botanically, vary a great deal in habit and general appearance. This is the reason, if you are establishing an area of prairie on your own land, for its being important to acquire your stock, whether seed or plants, from the closest prairie area possible to where you live. Local provenance is equally important in other countries, as emphasized elsewhere.

That is not to say you should yourself collect material from the wild, but you must obtain it from a reputable nursery that is conscientious about such matters. 'Nursery-grown' can be advertised after only one year on the nursery, the original seed or plants having come from anywhere in the world. 'Nursery-propagated' is the term you should look out for. Cheap offers of rare or endangered species are suspect, because the supplier may have robbed them from the wild.

You are unlikely to obtain all the species that you need from a local superstore or similar supplier. Mail orders will supply you with the widest range available. To avoid disappointment, you should order early and specify when you want delivery, whether in autumn or in spring. Burrell's advice when choosing a seed mix is to include as many species as you can, since it is often difficult to add plants once the garden is established. 'The popular meadow-in-a-can products are not a viable option,' he writes, since many are 'composed of non-native, even invasive, species'.

GRASSES

Grasses, which are the principal component of prairie, have devices for coping with prolonged drought. The tall grasses, up to head height or taller, have roots going 1.5–7.1m/5–23ft deep. Additionally, they often have a network of surface-feeding roots, which take advantage of any showers there may be in the course of a drought.

Grasses that grow in regions of hot summers tend to have two active seasons. The cool-season grasses start growing quite early in spring, flower in early summer but stop towards midsummer, restarting into growth in late summer and autumn. Warm-season grasses dominate most prairies. They begin growth in May or June, and flower in late summer. But there are cool-season grasses among them as a minor component, able to grow in dry or moist prairies. Some of the most ornamental cool-season grasses are found in the dry prairies.

For well-drained limestone or calcium-rich soil, sideoats, *Bouteloua curtipendula*, is a relatively low grass, growing in small clumps or as solitary stems up to 90cm/3ft.

The bright orange stamens of the summer flowers are followed by purplish seed heads. Other grasses with dense enough root systems to survive drought include porcupine grass (*Stipa spartea*) and prairie dropseed (*Sporobolus heterolepis*).

Little bluestem, *Schizachyrium scoparium* (60–120cm/2–4ft), is little only in comparison to big bluestem, its relative. It occurs as major ground cover in prairies ranging from extremely dry to mesic, on both acid and alkaline soil. Beth Chatto and I saw it in Minnesota and she now grows this clump-forming grass in her gravel garden in drought-prone East Anglia to erupt from low mat-forming plants. It is feathery with blue-green leaves that turn foxy-red in autumn, to combine in a dry prairie with *Artemisia ludoviciana* and asclepias and in less dry conditions (excessively moist does not suit it) with Michaelmas daisies and goldenrods. Cultivars have been bred to increase the blue of the leaves and the autumn colours.

Big bluestem, *Andropogon gerardii*, is a typical component of the original tallgrass prairie that ranged from Ohio to Colorado, from wet to dry. It grows in tufted bunches, the stalks usually from 90–180cm/3–6ft tall but reaching even higher where there is ample soil moisture in the warm-weather months. It has blue summertime foliage and bronze autumn colour. Selections have been made for specific climates and soils. At the New York Botanic Garden big bluestem blends with panicums and Indian grass, *Sorghastrum nutans* (90cm/3ft), as a setting for yellow solidagos and cup

Autumn-flowering switch grass (or panic grass), *Panicum virgatum*, makes a pinkish haze, both with its mass of tiny flowers and its foliage, which remains attractive all winter. There are many cultivated named varieties, of which I grow two at Dixter, 'Rubrum', seen here, and 'Hänse Herms'.

plant, *Silphium perfoliatum*, and the pinks of *Monarda fistulosa* and *Eupatorium maculatum* Atropurpureum Group.

Also for wet soil, there is switch grass (or panic grass), *Panicum virgatum* (1–2.4m/3¼–8ft, depending on conditions), which is drought-tolerant once established. It has an upright growth habit and good autumn colour. The airy flower panicles are often pink or red-tinted when first opening in late summer. It is worth researching the increasing array of selections, such as 'Prairie Sky' (blue leaves) and 'Shenandoah' (wine-coloured autumn foliage). Prairie cordgrass, *Spartina pectinata* (up to 3m/10ft), a spreading grass with graceful arching leaves, is native to wet prairies throughout the northern USA. It will tolerate average or dry soils, where its spreading is much reduced. 'Aureomarginata' is a gold-margined cultivar which I grow as a handsome but rather invasive border perennial.

THE FORBS

The forbs among the grasses have a long season, the small ones starting into flower quite early, while light is still reaching close down to ground level, before the grasses become active. The last of them to flower, golden-rods, sunflowers, veronicastrums (flowering July to early August), as also asters, are nearly as tall as the grasses themselves. On the dry prairies, the bloom has an early summer and an autumn peak, while on moist and dry prairies, blooming is more even throughout the growing season.

EARLY FLOWERS

Plants to grow in the prairie for early interest among emerging cold-season grasses and forbs, include several *Ranunculaceae*. *Pulsatilla patens*, the eastern pasque flower, is one of the first and most spectacular spring flowers of the northern tallgrass region, with pale lavender sepals followed by feathery seed heads. Another beauty of the buttercup family is the meadow anemone, *Anemone canadensis* (up to 60cm/2ft) – single pure white flowers on long, hairy stems – which appears in late spring. I have seen it in the wild, near Chicago, making delightful colonies beside a sluggish river. Marsh marigold, *Caltha palustris*, is found in wet, marshy areas through the northern tallgrass region, usually where competition from other vegetation is minimal. For dry soil there is the early buttercup, *Ranunculus fascicularis*, which really is early, and the later spring- and early-summer-flowering prairie buttercup, *R. rhomboideus* (both less than 30cm/1ft tall).

Of the *Primulaceae* shooting star, *Dodecatheon meadia*, grows up to 60cm/2ft from a cluster of basal leaves. The flowers produced in early June dangle, the reflexed petals being about 2.5cm/1in long, their colour ranging from light to dark pink and including a white form, *D. meadia* f. *album*.

The columbine – *Aquilegia canadensis* (red) and *A. formosa* (coral) – is not classified as a prairie plant but is easy to establish in a garden scheme for short-lived colour in July or August in combination with long-flowering prairie phlox, *Phlox pilosa*.

The camassias native to the tallgrass prairies are both mauve-flowered: the wild hyacinth, *Camassia scilloides* (about 30cm/1ft), is common in mesic and dry prairies through the eastern tallgrass region. The closely related prairie hyacinth, *C. angusta*, has taller stems and flowers later, in late spring to early summer, in prairies from Oklahoma to Indiana.

Lilies of the fertile American prairies include the prairie lily, *Lilium philadelphicum* (of moist prairies in the northern tallgrass region), with vivid orange flowers in early summer on stems up to 90cm/3ft tall. The only other true lily of the prairies, which also starts to flower in early summer, is the Michigan or Turk's cap lily (*L. michiganense*), sometimes confused with *L. superbum*. Prairie onion, *Allium stellatum* (45cm/1½in) is later, but as it is a member of the lily family, I include it here. Like those of other bulbous alliums, the leaves are basal and scruffy by the time the starry magenta flowers open in midsummer. It needs well-drained soil. I have already described clump-forming *A. cernuum* (page 172).

One of the most beautiful prairie perennials is false indigo, *Baptisia australis* ('australis' means 'of the south', nothing to do with Australia). It resembles a greatly refined lupin, the flowers well spaced on the spike and of a rich indigo blue in early summer. It has a deep, tough, gnarled taproot, which resents disturbance. Seed – soaked overnight in hot water before sowing – is the obvious method of increase, but the fresh young foliage is palatable to slugs in Britain.

LEGUMES

Unlike meadows, prairies are typically made up of a high percentage of legumes (*Papilionaceae*), whose root nodules fix nitrogen into the soil and boost the productivity of the grasses. Most, like the baptisias, finish flowering by the end of July. Blue false indigo, *Baptisia australis*, has obvious affinities with lupins, *Lupinus perennis*. The flower spikes (white on *B. alba)* are borne in early summer at 60–90cm/2–3ft, the plant making considerable vegetative growth afterwards. The flowers are well spaced, instead of being in whorls like lupins. The elongated seed pods are attractive, turning charcoal grey when ripe. The inflorescence of the somewhat taller *B. alba* var. *macrophylla* (syn. *B. lactea*) is a sparsely but widely spaced candelabrum of white-flowered spikes.

One of the most attractive legumes is leadplant or prairie shoestring, *Amorpha canescens* (up to 90cm/3ft tall and one of the few woody prairie plants). This has silvery green leaves and the lavender flowers are in dense florets. The vetches, *Astragalus*, tend to look weedier. Milk vetch, *A. canadensis* (up to 1.5m/5ft), with creamy white flowers, will tolerate all soils from dry to wet. Purple prairie clover, *Dalea purpurea* (60cm/2ft), has bright purple-magenta spikes and protruding orange stamens, forming a wreath that climbs the spike as flowering progresses. This and the white version, *D. candida*, grow in dry to mesic prairies throughout the tallgrass regions.

Goat's rue, *Galega officinalis* (1.5m/5ft), from Eurasia, has weak stems and, in borders needs support, which it gets in a prairie. The spikes of pea flowers are mauve typically and the flowering period, of three to four weeks, is centred on July.

COMPOSITES

Many of the showiest prairie ingredients belong to the huge family of *Asteraceae*, long known as *Compositae*. Among them are genera such as *Aster*, *Rudbeckia* and *Vernonia* that colonize North American meadows.

In the British climate the ironweeds (so called for their tough, rigid stems) flower almost too late in autumn. *Vernonia arkansana* (2m/6ft), the best-known species, is a rich shade of purple. Common ironweed, *V. fasciculata* (up to 1.2m/4ft), which grows in most of the tallgrass regions, often with prairie cord grass, is a brighter purple.

Sneezeweed, *Helenium autumnale* (up to 1.5m/5ft), is a stalwart of the old herbaceous border with a preference for damp soils. July to September is the season for its yellow flowers, which are yellow, brown and bronze in some of the cultivars, all of which are more suitable in relaxed plantings than in the cut and thrust of a true prairie. Purple-headed sneezeweed, *H. flexuosum*, is a true prairie plant, often found in acid soils through the southern tallgrass region and introduced as a weed northward. At 90cm/3ft, it is a good bit shorter than *H. autumnale* and the flower heads (yellow rays surrounding a purple brown disc) are in a flattish cluster.

Perennial sunflowers, *Helianthus*, are found throughout the tallgrass regions. Jerusalem artichoke, *H. tuberosus*, is a well-known thug which, at 2.4m/8ft, casts a lot of shade on lower plants growing alongside. *H. maximiliani* is equally tall and coarse and, also for damp soil, there is the sawtooth sunflower, *H. grosseserratus*, which has many large, narrow, alternate leaves along the stem, with a few teeth along the edges and hairy undersides. For drier soils, western sunflower, *H. occidentalis* (60–90cm/2–3ft), is altogether less gross, as is willow-leaved sunflower, *H. salicifolius*, with drooping leaves densely arrayed to form columns of foliage. In my borders it flowers late in September, in an open panicle of small daisies. In winter the stems remain interesting, taking on the greyish tones of oxidized lead.

Yellow daisies, of which there are many, compose an important element in the overall tapestry of any 'wild' planting. For those who like them tall, the sweet black-eyed Susan, *Rudbeckia subtomentosa*, grows up to 1.8m/6ft, given plenty of rich living and moisture. The yellow coneflower, *Ratibida pinnata*, is shorter (90–120cm/3–4ft), with rays bent back, almost parallel with the flower stem, surrounding a bullet-shaped disc.

The greyhead coneflower, *Ratibida pinnata*, has yellow petals bent back almost parallel with the stem and a prominent centre that turns from brown to grey in autumn. James Hitchmough succeeds with it but, grown from seed brought back from Minnesota, it petered out in my prairie, probably because the turf is dense and the plant is susceptible to slug damage.

I brought seed back from Minnesota. After growing it on for a year till we had large plants, we planted them in our prairie, where they flowered but were gone by the next year. The prairie coneflower, *R. columnifera*, native to the western edge of the tallgrass region, is similar, except that the central disc is longer.

The silphiums are yet another yellow, summer-flowering daisy. All are coarse-growing, moisture-loving prairie plants and among the tallest and largest-leaved. Cup plant, *Silphium perfoliatum* (1.8–2.4m/6–8ft tall, depending on the fertility of the soil), has abundant foliage which is opposite on square, flowering stems, and encloses them at the base, forming a cup. The daisies are only 7.5–10cm/3–4in across, but make a good display over a longish period. Prairie

dock, *S. terabinthinaceum*, makes a few, very large, broadly elliptical basal leaves. The flowering stems soar, almost leafless, to 1.8m/6ft or more and branch at the summit to quite small daisies. Compass plant, *S. laciniatum*, is similar but with cut leaves and green-bracted flower heads 6–11cm/2½–4½in wide.

As border plants, the goldenrods, *Solidago* spp., are generally despised because of their brash yellow colouring, but their bright, tubular flowers, borne in terminal spraying panicles, are effective as a prairie ingredient, with *Echinacea purpurea*, asters and grasses. I have not succeeded in establishing them in the dense turf of my prairie.

With *Echinacea*, close relatives of *Rudbeckia*, at last we arrive at composites with colouring other than yellow. The purple coneflower, *E. purpurea* (90cm/3ft), the

Rudbeckias and echinaceas are some of the ingredients in the 45 × 7.5m/150 × 25ft garden behind James Hitchmough's house in Sheffield. After years of research into prairie plants for British climates at Sheffield University, Hitchmough and his partner, Amanda Stokes, created a self-sustaining garden based on traditional blocks of American native plants that are at their best from late July onwards.

species we grow in our borders, is an easily grown, stiffly upright perennial, with hairy stems and leaves, crowned with large daisies for many weeks from July to September. The central boss is a prominent feature, made the more so by vivid orange anthers. Many populations in the wild are escapes from cultivation. *E. pallida* flowers earlier and is less showy, the petals being pale purple and the boss purplish brown. *E. angustifolia* is shorter, with smaller ray flowers.

The asters with root systems suitable for prairie plantings include New England aster, *A. novae-angliae*, with foliage that is dull and graceless in a border but fine in our context. Sky-blue aster, *A. oolentangiensis*, has numerous spreading flower branches. For prairie border plantings, try small-flowered asters, like *A. ericoides* and *A. cordifolius*. *Boltonia asteroides* (1.2m/4ft), aptly named false aster, is very similar to the typical Michaelmas daisy, with white flowers in October. For dry prairies, there are yellow-flowered coreopsis, including *Coreopsis palmata* and *C. grandiflora*.

ABOVE Spike gayfeather or dense blazing star, *Liatris spicata*, is very popular with butterflies. It is an unusual-looking composite with spikes of deep red-violet flowers that open from the top downwards in late summer until the first frost. It is often short-lived.

OPPOSITE, TOP The bristly seed heads of purple coneflower, *Echinacea purpurea*, remain attractive in winter and provide food for birds.

OPPOSITE, BOTTOM At the end of October, Hitchmough's urban garden is a haze of late-flowering *Rudbeckia fulgida* var. *deamii* and milky lilac *Aster laevis*. The bright-pink stems of burning bush, *Euonymus alatus*, will extend interest into late autumn and winter.

The eupatoriums are flat- or dome-shaped composites in which the display is made by the fluffy tubular florets and stamens. There are no rays. Joe Pye weed, *Eupatorium purpureum* (2.4m/8ft), is the best known and a good plant but somewhat superseded in our borders by *E. maculatum* Atropurpureum Group (1.8m/6ft), which is a slightly more intense shade of purple. With *Miscanthus*, pampas grass and late sunflowers, it creates a magnificent scene in autumn. There is a white form, *E. purpureum* subsp. *maculatum* 'Album'.

Prairie blazing star, *Liatris pycnostachya* (1.5m/5ft), is my favourite among this genus of prairie and savanna plants. The flowers consist of bright bluish-magenta tubular florets gathered in clusters along a spike, above stems that are liberally clothed in linear leaves. I have seen as many as five tortoiseshell butterflies supping simultaneously on one spike. A feature in liatris is that the spikes open – and die – from the top downwards. Dense blazing star *L. spicata* (60cm/2ft) is a shorter, sturdier plant with much shorter spikes. Both need moist soil and are reliably perennial, with tuberous rhizomes which can be divided. However, I have failed to establish them in my prairie – they lasted for just a couple of years. There are several blazing stars for drier soil, the best being *L. punctata*.

Yarrow, *Achillea millefolium*, is as ubiquitous in North America as it is in Europe (see pages 126 and 127) and is found along roadsides and in prairies throughout the tallgrass regions.

ASCLEPIAS

Most genera in this family (*Asclepiadaceae*) are North American. The butterfly milkweed, *Asclepia tuberosa* (60cm/2ft), is a hardy clump-forming prairie plant, which flowers in early summer in its native land but waits until late summer to perform in the cooler British climate, where temperatures are rarely sufficiently hot to please it altogether. Seedlings, Hitchmough has discovered, are susceptible to slug damage. The flowers borne in umbels are a wonderful shade of rich orange. It makes a splendid contrast to blue flowers, such as blue sage, *Salvia azurea.* Swamp milkweed, *Asclepia incarnata*, with reddish-pink flowers, is the one for saturated soils. Spider milkweed, *A. viridis*, is native to well-drained or rocky prairies.

ERYNGIUM

Rattlesnake master, *Eryngium yuccifolium*, is holding its own in my own prairie, reaching 75cm/2½ft (it grows quite a bit taller in the USA), although it does not go so far as to self-sow, which it does, too much, in its native land. The common name is said to derive from brewing

Eryngium yuccifolium has a thick taproot that needs space to develop. My seed came from the American Midwest and the resulting plants have held their own well under very competitive conditions, although they have not reached the stage of self-sowing that would fully endorse their at-home-ness in my prairie.

its root as an antidote to rattlesnake poison. With yucca-like leaves, it makes a branching candelabrum of small, pale green inflorescences that look handsome with big bluestem and flowering spurge, *Euphorbia corollata.*

MONARDA

Bee balm, Oswego tea, bergamot, horsemint, monarda, whatever you like to call this American genus, it needs rich, moisture-retaining and deeply cultivated soil. The plants overwinter on a dense network of stolons carrying rosettes of small leaves. In the border these are apt to kill themselves out by overcrowding (we surface mulch ours with old potting soil in spring, just as the young shoots are lengthening). In a prairie, competition from grasses is needed to control their spread.

Wild monarda, *M. fistulosa*, has the strong lemony aroma common to all monardas. The mauve-pink hooded flowers are densely arranged in two or three whorls towards the top of 1m/3¼ft (or more – up to 1.5m/5ft in the tallgrass regions) square stems. There are some beautiful cultivars available, in shades of red, purple, mauve, pink and white, all of which would be worth experimenting with in a prairie, so long as the soil never dries out. The dark circular seed heads are an asset in winter. *M. didyma* is an attractive species, typically with red flowers.

VERBENAS

Blue vervain, *Verbena hastata*, has multiple mauve flower spikes. I prefer the colour of *V. hastata* f. *rosea*, which grows to 1.2m/4ft in my garden and, I'm told, up to 1.8m/6ft in wet areas of the tallgrass prairies.

VERONICASTRUM

Culver's root, *Veronicastrum virginicum*, has done outstandingly well in my prairie, reaching 1m/3¼ft. At home, in the western-most tallgrass region and along roadsides, it can reach 1.5m/5ft, combining well with ironweed, *Vernonia* spp., and big bluestem. The culver's root is boldly upright and has presence. Small veronica-like flowers are held in dense clusters of tapering spikes at the tops of the branches, with pointed green sepals and a tubular white corolla. I have never seen a blue, but you can order *Veronicastrum virginicum* 'Album' to be certain that white is what you get; *V. virginicum* f. *roseum* is pale pink and there are cultivars offering other shades. The serpent-like seed heads are wild and interesting in winter.

11. The Wider Landscape

IRREMEDIABLE HARM TO THE CONTENTS of a meadow can be achieved in a few hours. With increasing awareness of the environment and of issues of biodiversity, there is considerable pressure to conserve the meadows that remain. The political climate for creating and restoring species-rich grassland has been greatly strengthened with farmers being forced to find new economic uses for their land. Ploughing is no longer permitted on uncultivated land, but the widespread use of fertilizers continues. So does the practice of two or three annual green cuts for silage in the place of hay. Meadows containing a varied tapestry of flowers that provide food for insects and ground-nesting sites for birds such as skylarks continue to be 'improved', all too quickly becoming green fields with as few as three or four species.

Governments are committed to helping farmers diversify, balancing the requirements of modern farming with the objectives of conservation. Diversification includes the creation and long-term maintenance of meadows. However, relic grasslands are still under threat, either from change of use or from lack of management. And it has to be said that many people attracted to the romantic concept of a flowery mead are unpractical when it comes to the nitty-gritty of restoration, whether meadows or wildflower grassland on roadsides and in urban areas. Such plantings are promoted as environmentally sound and

A striking contrast in the Derbyshire Peak District between fertilized fields that are cut for silage two to three times a year and contain very few species, and an untreated hay meadow full of many wildflowers. Low nutrient levels suit rough hawkbit, *Leontodon hispidus*, and quaking grass, *Briza media*. Common cat's ear, *Hypochaeris radicata*, and the little white-flowered common mouse ear, *Cerastium fontanum*, rather like a miniature stitchwort, are abundant.

cost-saving alternatives to conventional ground covers, but there is often more work and attention required than they had bargained for. They cut the odd corner, botch the job and are disillusioned. I think you need to start out with a good measure of humility and not to be resentful of a spot of hard work and intelligent vigilance. You need to realize that you are dealing with living things in a range of habitats, so every case is unique and will require a slightly different solution that suits local conditions.

LOCAL INITIATIVES

For nature conservation purposes, the emphasis on using seed of local provenance has become increasingly important and is an aspect of initiatives to encourage the management of existing wildflower meadows. Seed from another country or even from another region, although of the desired species, is likely to be of quite different habit from the local product. With ox-eye daisies, for instance, many of the highway stands that we see ablaze in solid white, in early June, are grown from a vigorous strain of imported seed. The plants may not last for more than two or three years. Seed from a local strain produces durable plants of less vigour. Colston Burrell offers the example of switchgrass, *Panicum virgatum*, from Nebraska being taller and coarser than the same species from Minnesota. 'When planted in Minnesota,' he writes, 'the Nebraska plants look out of place and may cross-pollinate with local plants, thereby compromising the genetic integrity and changing the appearance of the Minnesota plants.' The grasses used for ecological restoration in Britain (paid for out of public funds) are cultivated varieties, often from the Continent, New Zealand or North America. They may have been developed for their vigour or for their suitability for football pitches.

The principle of using local wildflower seed that is genetically adapted to the area needs to be followed worldwide. In France rigorous measures are applied to ensure the genetic conservation of threatened wild plants, and in Scandinavia the Nordic Council of Ministers has established a Nordic Gene Bank to conserve both native and introduced genetic resources in Scandinavia and Iceland, especially populations of native forage grasses and clovers.

A good example of what can be achieved to increase the availability of local seed is current in the area where I live. The Weald Meadows Initiative (WMI), already cited in Chapter 4 on the creation of grasslands, generates income for owners of ancient wildflower meadows. It harvests (in suitably dry weather) seed, cleans, dries and packages it and sells it to farmers and nature-loving amateurs who want to enhance existing grassland or create new areas. The initiative also monitors the seed and provides the much-needed advice on how to make these projects a success. Because the initial stages are not simple and certain rules need to be strictly followed, there are a good many failures, but there is no need for these if the right things are done at the right time. Initial preparation and follow-up management are essential for sowing success. Many failures are caused by small lapses in care and inappropriate timings of operations.

Timings should follow the typical annual cycle of a meadow:

Early spring Grazing by stock continues. Ground-nesting birds and mammals establish their territories.
Late spring to midsummer No grazing. A rich and varied tapestry of flowers provides food for insects and ground-nesting sites for birds such as skylarks.
Midsummer The grasses ripen and the meadow is cut. The dried hay is baled, carried and stored for winter feed.
Late summer to autumn The grass soon greens up and dominates in late summer. Grazing starts again and continues through autumn. If the meadow needs reseeding, turning cattle on to the land until late September will open up the ground sufficiently to allow successful broadcasting of seed (of local provenance).
Winter Livestock may remain on the meadow. The summer's hay is fed to it.
Late winter Small amounts of well-rotted farmyard manure from the overwintering barns is spread on the meadows.

With this cycle in mind, the WMI offers guidelines on enhancing existing grasslands:

- At no stage in any meadow management should there be so much as a sniff of artificial fertilizer.
- A weed wipe or carefully controlled individual spraying with glyphosate should be applied to any docks, nettles, thistles and similar perennial weeds.
- The sward must be cut short and harrowed to expose bare soil, then sown, late summer to early winter, at 3–5kg per hectare/2 –4½lb per acre, by hand or machine.
- The seed can either be broadcast overall or in bare patches.
- Roll after sowing (depending on soil conditions).
- In the first year cutting must be regular (to prevent flowering), and the cuttings removed.
- Grazing is the alternative to cutting. The WMI is investigating the feasibility of a grazing and cutting of small species-rich sites, thus conserving the traditional management that is so important for meadows. With sufficient beasts, grazing might be needed for only two or three days, but fencing may be a problem, although nowadays there is transportable electric fencing.
- No flowering or seeding should be allowed in the first year.
- In year two summer hay cut after plants have seeded. Thereafter either take a second cut or allow the grass to regrow and graze into early winter. A light harrow in mid-autumn may be beneficial, as it will aid germination of any remaining seed.
- Reduce the coarseness and vigour of the sward by introducing hay rattle (*Rhinanthus minor*).
- Noxious weed seeds are bound to creep in from time to time and must be dealt with on an individual basis.

URBAN VEGETATION

ABOVE Weld, *Reseda luteola*, an ancient dye plant, is a stately biennial with feathery greenish-yellow flowers. In fields where Landlife grows meadow plants for the production of seed for reclamation schemes, relics of weld persist when the ground is disturbed.
OPPOSITE In an urban reclamation project in the Liverpool area, the rather late-flowering (June–July) meadow cranesbill, *Geranium pratense*, is handsome amid a haze of scrambling hedge bedstraw, *Galium mollugo*. Richard Scott of Landlife values the meadow cranesbill as a robust plant of grassland, competing even with Yorkshire fog (*Holcus lanatus*), although it takes a few years to flower from seed.

The green areas in many urban landscapes are extremely dreary – the grass is either gang-mown, perhaps sixteen times a year, or planted with labour-saving shrubs, which are predictably dull in themselves, but keep the weeds down. At least they are green, you may say, rather than built over, but always provided that the cost is comparable with current practices, the introduction of one form of meadow or another, temporary or permanent, is clearly more attractive to people and wildlife.

Public spaces present particular problems: a diverse group of people having different views on the subject gives such actions a high profile. The best way to keep everyone happy is to consult and involve both groups and key individuals (who are often the noisiest). Visible success, without much waiting, is essential. In this respect, the long-term establishment of grassy meadows can present problems. In the first year nothing much happens. On your own territory you might shrug your shoulders and say, 'I can wait.' But in urban public spaces, this is unacceptable. The grass meadow concept is anyway inhibited by lack of suitable machinery for cutting long grass and by lack of any method for its ready disposal. Transporting it to a landfill site is prohibitively expensive. Landlife claims that chopping the herbage very finely by flailing several times at a go, and leaving it *in situ*, does not materially increase soil fertility. A spring and a late summer cut will do the trick.

The quick returns from meadow plantings of annuals and biennials are an easier sell to the public. Richard Scott took me to see an urban reclamation project in the Liverpool area for which Landlife supplies seed, as required, and advice. Pickering's Pasture, a large meadow overlooking the Mersey and popular with the public, is administered by the enlightened Halton Borough Council. Until the 1950s the area was grazed by cattle and home to wading birds and estuary plants. For the next thirty years, however, a mountain of refuse built up on the salt marsh. In the 1980s the area was covered with industrial fuel ash (inert but

of a high pH), thereby recycling a waste product; then with a thick layer of clay, and finally with soil. The site was divided into three, each area concentrating on a different season. Perennial weeds like creeping thistle, which would be liable to take over large areas, are spot-treated with glysophate (Roundup), except in a control area, for comparison. The grass is cut just once a year, in August, and as its growth is comparatively feeble, it can be chopped up fine and left *in situ* without noticeably affecting fertility. This reduces management costs dramatically. Where the turf has become dense, it is chain-harrowed (equivalent to grazing cattle on it) to open up gaps and can then be oversown with a limited selection of perennials, such as field buttercup (*Ranunculus acris*) and ox-eye daisies, but few others would establish. I was especially taken by a sizeable area of chicory, *Cichorium intybus*, 2m/7ft tall and full of its blue flowers. This is perennial, if conditions are to its liking. We have just planted some roots into our prairie area.

Landlife welcomes the introduction of colourful aliens, if they achieve a purpose. Feral lupins were a common sight in areas that were later built over, but they have been reintroduced to other areas around new

This jolly mix of directly sown annuals, devised by Nigel Dunnett of the University of Sheffield, is still in full flower in the middle of October. The 'meadow' includes yellow *Coreopsis tinctoria*, white *Ammi majus* and red poppies, and is set in the heart of Sheffield where the public passes through and appreciates the colourful surroundings. Children in particular find the plants magnetically attractive and gather large bunches. This is a form of interaction (Dunnett calls it 'positive vandalism') which is welcomed in a setting where real contact with nature has been all but eliminated.

housing, where their cheerful contribution is welcomed. That is a quick-developing perennial such as Nigel Dunnett includes in his experiments with biennials and quick-maturing though short-lived perennials. These include verbascums, teasels, *Digitalis purpurea* (foxglove), *Hesperis matronalis* (sweet rocket), *Leucanthemum vulgare* (ox-eye daisy) and *Salvia sclarea* var. *turkestanica*. They perform in their second year and the short-lived perennials make a further display in the third. To achieve a 'return' in the first year, you can include 10 per cent of annuals in the seed mix, choosing species of thin texture, like *Linaria maroccana* and *Coreopsis tinctoria*, which will not unduly shade out the developing biennials. Even so, they do have an inhibiting effect.

ROADSIDES

Large, deliberately sown but short-lived swathes of ox-eye daisies, *Leucanthemum vulgare*, are particularly successful on newly disturbed roadside banks. Opening in late May, they are at their best through June.

A very large proportion of the relatively permanent meadow areas left to us is included on roadsides. You must be aware of them as you sweep past them in a car. Stop! Stop! you shriek at the alarmed driver, but there is no possibility (in most cases) of stopping, because of other traffic or the lack of a handy lay-by or even because the driver is unsympathetic to the notion of interrupting the journey for the footling reason of an interesting plant, or plants, having been spotted.

If we want to save significant plants on roadsides, we must report sightings to local wildlife pressure groups, which have schemes for recording interesting species and habitats (mostly in the form of a database). On the road into Hastings near where I live in East Sussex, I was delighted to spot a colony of common spotted orchids but, on closer inspection – the driver pulled in – saw they were being overtaken by brambles. Before giving a verge special treatment, the authorities responsible for roadsides often wait until they have received notification. Wildlife verges that have been identified, and are subsequently protected and managed, set an excellent example of what is possible, but constitute only a tiny proportion of all those that need care and attention. Miriam Rothschild has been a leading proponent of native wildflowers on roadsides and has established flourishing colonies of her mixes selected from recreated ancient wildflower meadows.

Large stands of biennial *Verbascum speciosum* are commonplace on roadsides and in abandoned fields in Hungary where an open-textured sward allows them to form colonies.

There are also examples of entirely new grasslands created alongside new roads by teams of horticulturists, ecologists, environmentalists and landscape architects working with the engineers towards improving bio-diversity. I think everyone who has driven along the Brighton Bypass in Sussex must admire the contouring either side of the tunnel through which it runs. Geologically, this is almost raw chalk. Topsoil only 2–10cm/¾–4in deep was laid over it in the form of Geotextile, which is a honey-combed material. Wildflowers were planted on the terraces and a wildflower mixture was sown in with grasses. The initial specification for grasses included 60 per cent of ryegrass, which was what the engineers wanted because it germinates quickly and lessens the danger of soil washing away. Henry Penner, Senior Environmental Adviser, got this down to 10 per cent, as he wanted finer grasses that would be far more suitable in the long term. But they are slow to germinate, which caused the engineers to fret. For the first three years the grass was cut twice a year. It now hasn't been cut at all for five years. Management is always a balancing act – do you cut regularly or go in once in a while to face a bigger job? Which is cheaper? All of this has to be monitored and worked out. Once the engineers have exercised their prerogatives, full emphasis on biodiversity can hold sway.

Roadside verges are defined as the area (of varying width but a minimum of 1m/3¼ft) between the road itself and a boundary behind it, be it wall, hedge, fence or other barrier. They comprise miles and miles (and even more kilometres) of potential nature reserve, forming corridors along which wildlife, whether animal or vegetable, can connect and be transported without crossing unsuitable territory. Much consists of scrub, which is good for some forms of wildlife, but grassland provides habitats for a far greater range of species.

Although roadsides are relatively undisturbed by humanity, there are yet many factors contributing to the disturbance or extinction of their plant communities, including the addition of out-of-character daffodils or other alien plants. Inappropriate cutting regimes, or lack of any, may lead to invasion by all sorts of coarse herbage. Roadsides also suffer from dumping and spoil tipping, car and lorry parking, picnicking and peeing, machinery used for hedge-cutting and, particularly in North America, the clearing of vegetation for the benefit of billboards. Yet further factors, such as pollution by spillage from vehicles, salt spray run-off from roads and herbicide drift, impinge on the sort of meadow communities we should like to foster. Chemicals for growth and weed control are used minimally, nowadays, largely because of their expense! We often do the right thing for the wrong reason, but public opinion is here again highly influential.

Where policies are set up, there is the opportunity for manipulating land for the benefit of certain ecotypes which would otherwise be extinguished. One dynamic person to get things going can achieve a great deal and the message is passed on. One such mover and shaker in the USA was Lady Bird Johnson. In 1982 she founded a research centre in Austin, Texas, dedicated to 'the study, preservation, and re-establishment of native plants in public and private landscapes'. In 1987 Mrs Johnson was instrumental in assuring the inclusion of a provision into a Federal Highway Act that required a

Lomatium martindalei is happiest on the dry, sunny side of the Columbia Gorge in Portland, Oregon. As trucks thundered by, we saw it flowering with blue *Dichelostemma congestum* (syn. *Brodiaea congesta*), tufted vetch (*Vicia cracca*) and sweeps of California poppy, *Eschscholzia californica*, all flourishing in the open-textured sward.

percentage of funds to be spent on revegetation with wildflowers, inaugurating a long tradition of seeding road-verges and medians with natives (especially each state's official wildflower). This caused a surge in highway beautification. Mrs Johnson was continuing a movement that began in the 1930s when women's garden clubs of Texas were active in conserving their roadsides, at a time when the verges were ploughed with mule-drawn mowers. This had led to the statutory mandate not to cut the massed stands of Texas blue-bonnets until they set seed.

A scattering of ox-eye daisies among yellow and orange hawkweeds growing in abundance makes a dramatic combination on a roadside in Vermont. I have failed to establish the orange hawkweed in the dense turf at Dixter, although it thrives to the point of being invasive in our borders.

Corridors of unspoilt prairie flora and fauna survive along highways in the USA but, as in Britain and Europe, roadsides are still sprayed with herbicide, unless attention has been drawn to some extremely rare and glamorous species (like an orchid) that is growing in it. There is often misunderstanding of how even verges from which herbicide sprays are excluded should be treated. Highways departments continue to use non-native grasses because the seed is cheaper. Only a small percentage of federal money is spent on highways, with few guidelines on management techniques. Swathes of colourful, usually exotic, plants are installed to catch the eye. I was disgusted on one Georgia road along which I was driven by a huge planting of a single variety of canna. It had been decimated by some kind of leaf-eating pest. This is the sort of problem that a monoculture immediately leads to. But in any case, the inappropriateness of cannas in a semi-rural context was glaring. Cosmos are colourful annuals and popular with the public, but appropriate, *en masse*, only to gardening in a public park. This bears no relation to the kind of recreation of a prairie flora that the thoughtful treatment of an available roadside area could bring about. Again, this is where an informed public can make the difference. Those with eyes for native plant communities should report sightings to local pressure groups.

DYNAMIC WOODLAND EDGE

Natural meadows are usually bordered by woodland, the trees providing a backdrop and placing the meadow in context – reminding us that, given time and neglect, woody species take over and a meadow happily reverts to forest. Taking the broader view, the solidity of shrubs is clearly needed as a break, even in an otherwise open landscape, while shrubs may be needed as a transition from woodland to, say, meadow pure and simple. The edge also provides habitat for wildlife, especially mammals and birds.

The light stems, frequently pollarded, of *Salix alba* var. *vitellina* 'Britzensis' (syn. *S. a.* 'Chermesina') need a dark background to highlight them in winter sunshine. This willow is vigorous enough to grow in dense meadow turf and at Dixter we grow it as a solitary specimen on the edge of the lower moat. Nigel Dunnett groups stooled willows and other suckering plants in his schemes for meadow edges.

Traditionally, landscape designers are often extremely unimaginative in their choice of dreary, no-trouble shrubs. Dr Nigel Dunnett, again of the Department of Landscape, University of Sheffield and again at Harlow Carr, is concerned with the broad aspects of landscaping, but clearly his work has a more intimate application for many gardeners, witness his article 'Coppice Gardening' in *The Garden*, March 1995.

Dunnett is demonstrating how shrubs of a certain type can be combined with a varied assortment of perennials, biennials and annuals, both native and exotic. His model is the coppiced woodland; the under-storey of woodland that is coppiced at intervals (in our own woods at Dixter, this may be a shorter interval of ten or twelve years with chestnut coppiced for fencing, to twenty or even twenty-five years for coppice mainly used as firewood). Beneath this woody canopy is a flora adapted to long intervals of twilight alternating with shorter intervals when light is abundant. Such plants are capable of ticking over in the darker years, without flowering abundantly if at all; for example, primroses, wood anemones, bluebells and early-purple orchids. Bluebells are never so abundant as after coppicing. I am here writing of southern Britain. In Scotland, where the light is softer, they are seen in far more open situations, as are primroses, their growth often alternating in the same year, with that of bracken.

Alternatively, in the coppicing rotation, the ground flora passes the dark years as dormant seed, which germinates when the light returns. Such are foxgloves and red campion (*Silene dioica*). During these light periods they have the advantage of being on the spot and ready to take advantage of the situation. Their potential rivals, plants which need open, sunny conditions at all times, are simply not there to move in. The foxgloves of this world cannot exist in dense turf, but in the above scenario they enjoy periods when they are, so to speak, cock-of-the-walk.

Dunnett has devised a more ornamental and varied community, wherein the shrubs are important protagonists. Overall maintenance, once the stage is set, is (he claims) neither burdensome nor costly. The shrubs are such as can be stooled (that is, cut back) to ground level at frequent intervals, say every three to five years. So, at ground level, there is quite a variation between very light and fairly shaded in the course of those years.

SHRUBS

The shrubs must be planted in weed-free ground, and spaced 2m/7ft apart. Particularly suitable for stooling are the sumacs, *Rhus glabra, R. typhina*, hybrids between them and cut-leaved cultivars. They have a suckering habit, which is an advantage in the context, and their leaves colour up in autumn. The related smoke bushes, *Cotinus*, are also good, whether green- or purple-leaved, and they too have excellent autumn colour. Amelanchiers are good for stooling, flowering well, with impressive autumn colour and, often, a suckering habit.

Other examples will readily suggest themselves. Many willows are seen at their best when stooled; for instance, the red-stemmed kinds and those with silvery foliage, like *Salix alba* var. *sericea*. One that I should particularly like to try is the American species, *S. exigua*, which has the most silvered leaves of all, combined with a naturally suckering habit. Stooled *Acer* in a range of species, even those normally making trees, are especially suitable and effective, as are cultivars of elder, *Sambucus nigra*, grown principally for their coloured or variegated foliage; additionally, they would flower towards the end of their five-year cycle.

As suckering shrubs, the spiraea-like sorbarias, with pinnate foliage and creamy flower panicles terminating their young growth, have proved a success. Some roses sucker freely, if on their own roots, and can be cut to the ground without demurring. *Rosa rugosa* is noteworthy and I would suggest *R. virginiana*, comely at all times of the year, for stem colour in winter, leaf colour in autumn, charming summer flowers and crops of red hips. Stooling rejuvenates it a treat. Sea buckthorn, *Hippophae rhamnoides*, grows in the wild on stony ground close to the sea but it is adaptable, with silver-grey foliage and crops, from female plants (both sexes must be present), of luminous orange berries. That suckers.

THE HERBACEOUS LAYER

The bare ground around each newly planted shrub offers an obvious opportunity for establishing the herbaceous layer, whether planted or grown from seed. Dunnett recommends limiting the number of different herbaceous species for more telling effect and I would strongly endorse that. With an eye on impact (also on management) he suggests concentrating either on early plantings, with a March to July season, such as you would be likely to find in a European coppice; or on late-season plantings, flowering June to October, which is more the North American pattern for woodland edges, roadsides and prairies.

For early season effect Dunnett's perennials, beneath *Amelanchier canadensis*, include: *Geranium sylvaticum* and *G. psilostemon; Digitalis ferruginea*, a perennial foxglove with coppery-yellow trumpets; doronicums, leopard's bane, *Doronicum orientale*, with yellow daisies; and *Brunnera macrophylla*, of the borage family, with blue flowers in spring. Species sown in include sweet rocket, *Hesperis matronalis*; biennial honesty or money flower, *Lunaria annua*; sorrel, *Rumex acetosa*; field buttercup, *Ranunculus acris*; and red campion, *Silene dioica*. Maintenance involves a late-summer cut of the herbage, removing the cut. Stooling the shrubs, has to be done as a separate exercise in winter.

An example of Dunnett's late-season plantings (which he clearly prefers) sees black-eyed Susan, *Rudbeckia fulgida* var. *deamii*, flowering dramatically among *Rhus typhina* and *R. glabra* 'Laciniata', the shrubs' leaves showing autumn colour. In this case, the cutting and removal of herbage is left until winter and can be simultaneous with the stooling of shrubs, when due. If there is competition from invasive grasses, the rudbeckia's vigour and flowering is much reduced. However, tussock-forming grasses that keep to themselves, whether native or exotic, are no problem to forbs.

The edge between woodland and meadow at Harlow Carr, Yorkshire, with the shrubby suckering stag's horn sumac, *Rhus typhina*, and black-eyed Susan, *Rudbeckia fulgida* var. *deamii*. The vigour of rudbeckias is greatly reduced if there is competition from aggressive grasses.

BIBLIOGRAPHY

BURRELL, C. Colston *A Gardener's Encyclopedia of Wild Flowers* Rodale, 1997

CAVALLO, Adolfo Salvatore *The Unicorn Tapestries* The Metropolitan Museum of Art and Harry N. Abrams, 1998

CHATTO, Beth *Beth Chatto's Woodland Garden* Cassell Illustrated/Sterling, 2002

DRUSE, Ken *The Natural Habitat Garden* Clarkson Potter, 1996

FELTWELL, John *Meadows: a history and natural history* Alan Sutton, 1992

FITTER, Richard and Alastair, and Marjorie BLAMEY *The Wild Flowers of Britain and Northern Europe* Collins, 1974

HOBHOUSE, Penelope *Gardens of Persia* Cassell Illustrated, 2003/Kales Press, 2004

– *Plants in Garden History* Pavilion, 1992

– *Natural Planting* Pavilion/Henry Holt, 1997

– *The story of Gardening* Dorling Kindersley, 2002

LADD, Doug *Tallgrass Prairie Wildflowers* Falcon Press (in cooperation with The Nature Conservancy), 1995

LEWIS, Pam *Making Wildflower Meadows* Frances Lincoln, 2003

LLOYD, Christopher *Christopher Lloyd's Garden Flowers* Cassell&Co/ Timber Press, 2000

LORD, Tony *Gardening at Sissinghurst* Frances Lincoln/Macmillan USA, 1995

MABEY, Richard *Flora Britannica* Sinclair-Stevenson, 1996

RICE, Graham *Discovering Annuals* Frances Lincoln/Timber Press, 1999

ROBINSON, William *The English Flower Garden* John Murray, 1905

SMITH, J. Robert with Beatrice S. SMITH *The Prairie Garden* The University of Wisconsin Press, 1980

STACE, Clive *New Flora of the British Isles* Cambridge University Press, 1997

THACKER, Christopher *A History of Gardens* Croom Helm/University of California Press, 1985

WASOWSKI, Sally *Gardening with Prairie Plants* University of Minnesota Press, 2002

ARTICLES AND PAMPHLETS

BISGROVE Richard and Paul HADLEY 'Gardening in the Global Greenhouse: Climate Change and Gardens', 2002

BUCKINGHAM, Helen, Jane CHAPMAN and Rebekah NEWMAN 'Meadows beyond the Millennium' Peak District National Park Authority, 1999

LICKORISH, Su, Grant LUSCOMBE and Richard SCOTT 'Wildflowers Work: a technical guide to creating and managing wildflower landscapes' Landlife, 1997

'Meadows and Meadow Gardening' *New England Wild Flower Society* Volume 5 Number 1

The Weald Meadows Initiative three-year report 1996–1999

WOUDSTRA, Jan, and James HITCHMOUGH 'The Enamelled Mead: history and practice of exotic perennials grown in grassy swards' *Landscape Research* Volume 25 Number 1, March, 2000

INDEX

Page numbers in **bold** refer to the illustrations

A

Acer, 226
Achillea, **64**
A. millefolium (yarrow), 59, 126, **127**, 212
A. ochroleuca, 70
aconite, winter SEE *Eranthis hiemalis*
adder's tongue fern SEE *Ophioglossum vulgatum*
agriculture, 14, 32–3, 56–7, 62–5, 70, 182, 195–6
Agrimonia (agrimony)
A. eupatoria, 147
A. procera (fragrant agrimony), 148
Agropyron repens SEE *Elymus repens*
Agrostemma githago (corn cockle), **181**, 184
Agrostis capillaris (common bent), 28, 77, 115, **116**, 117, **128**
A. stolonifera (creeping bent), 115, 183
Ajuga reptans (bugle), 153
Albertus Magnus, *De vegetabilibus*, 60
alien plants, 73, 223
Allium (onions), 11, 172
A. cernuum (nodding onion), 172, **172**, 205
A. oleraceum (field garlic), 172
A. paradoxum (few-flowered garlic), 172
A. roseum (rosy garlic), 172
A. sphaerocephalon (round-headed leek), 172
A. stellatum (prairie onion), 205
A. triquetrum (three cornered leek or garlic), 120, 172
A. vineale (crow garlic), 125
Alopecurus pratensis (meadow foxtail), 115, **115**
Alps, **62**, 113
Amelanchier, 226
A. canadensis, 226
Ammi majus (bishop's flower), 188, **189**, **220**
Ammophilia arenaria (marram grass), 18
Amorpha canescens (lead plant, prairie shoestring), 195, 205
Anacamptis morio (green-winged orchid), 81, **138–9**, 139–40
A. pyramidalis (pyramidal orchid), **102–3**, 138
Andropogon gerardii (big bluestem), 195, 203
Anemone apennina, 135, **91**
A. blanda, 135
A.b. 'White Splendour', 135
A. canadensis (meadow anemone), 136, 204
A. cylindrica (candle anemone, thimbleweed), 136
A. nemorosa (wood anemone), **76**, 79, 83, **83**, 135, **135**, 146, 225,
A.n. 'Lismore Blue', 135, **135**
A.n. 'Robinsoniana', 135
animals, 14, 35, 49, 55, 56, 66, 182
annual meadows, 44, 181–191
cornfield plants 183–6
early-flowering plants, 189–90
late-flowering plants, 190–91
long-flowering plants, 188–9
native plants, 187
sowing seeds, 183
tall plants, 191
weeds, 186–7
annual weeds, 99, 183, 201
annuals, as catch-crops, 95
Anthemis arvensis (corn chamomile), **180–1**, 184, 185
Anthoxanthum odoratum (sweet vernal grass), 115
Anthriscus sylvestris (cow parsley, Queen Anne's lace), **76–7**, **83**, 91, **91**, 150, **151**, 178
Anthyllis vulneraria (kidney vetch), 97
antirrhinums, 154
Aquilegia (columbine), 57, 59
A. canadensis, 204
A. formosa, 204
A. vulgaris, 136
Argemone mexicana (Mexican poppy), 189
Arrhenatherum elatius (false oat-grass), 100, **56**
Artemisia (wormwood)
A. ludoviciana, 203
A. pontica, **65**, 66
artichoke, Jerusalem SEE *Helianthus tuberosus*
Asclepias (milkweed), 201, 203, 212
A. incarnata (swamp milkweed), 119, 212
A. syriaca, **67**, 89
A. tuberosa (butterfly milkweed), 22, 212
A. viridis (spider milkweed), 212
Asparagus officinalis, 69
Aster, 119, 204, 206, 209, 210
A. novae-angliae (New England aster), 119, 210
A. oolentangiensis (sky-blue aster), 210
aster, false SEE *Boltonia asteroides*
Astragalus (vetches), 205
A. canadensis (milk vetch), 205
Astrantia, 151
A. major, **151**
A.m. 'Ruby Wedding', **151**
Atriplex hortensis (purple orach), 187, 190, 191
aubrieta, **78**
autumn crocuses, 106, 160, **160**
autumn lady's tresses SEE *Spiranthes spiralis*
Avena (oats), 181, 184, 196
Avon Gorge, 172

B

baby blue eyes SEE *Nemophila menziesii*
baby's breath SEE *Gypsophila elegans*
bachelor's buttons SEE *Centaurea cyanus*
badgers, 88, 141, 152
Balaton, Lake, **62**
Baptisia, 201, 205
B. alba, 205
B.a var. *macrophylla* 205
B. australis (blue false indigo), 201, 205, **205**
B. lactea, 205
barley SEE *Hordeum*
bartsia, red SEE *Odontites vernus*
basil, wild SEE *Clinopodium vulgare*
Beavis, Ian, 43
bedstraws SEE *Galium*
bee balm SEE *Monarda*
Beech Estate, 38–9
bees, 11, 14, 16, 19, 24, 41–3, **41**, 49, 120, 121, 131, 138, 144, 152, 158
bellflower, creeping SEE *Campanula rapunculoides*
bent SEE *Agrostis*
bergamot SEE *Monarda*
betony SEE *Stachys officinalis*
biennials, 193, 218, 221, 225
bindweed, 186
birds, 10, 14, 18, 105, 159, 167, 187, 195, 211, 215, 217, 218, 225

bird's-foot trefoil
SEE *Lotus corniculatus*
bishop's flower SEE *Ammi majus*
bison, 194, 196
bistort SEE *Persicaria bistorta*
Black Dog Reserve, Minnesota, 193, 196
black-eyed Susan SEE *Rudbeckia*
blanket, wildflower, 94
blazing star SEE *Liatris*
bluebell SEE *Hyacinthoides*
California SEE *Phacelia campanularia*
bluebonnets, Texas
SEE *Lupinus subcarnosus*
bluegrass SEE *Poa*
bluestem
big SEE *Andropogon gerardii*
little SEE *Schizachyrium scoparium*
Boccaccio, Giovanni, *Decameron*, 60
Boltonia asteroides (false aster), 195, 210
Borago officinalis (borage), **188**, 189
Botrychium lunaria (moonwort), 142
Botticelli, Sandro, *La Primavera*, 58, **59**
garden, at Dixter 74, 78, 79, 136, 139, 167, 173
Bouteloua curtipendula (sideoats), 202
Bowles, E.A., 159
bracken, 225
Braunton Burrows, Devon, 126
Brickwood, Dawn, 38–9
Brighton, East Sussex,
Bypass, 222
Parks, 8
Briza (quaking grass)
B. maxima, 191
B. media, 69, **116**, 117, **215**
Bromus (brome), 198
Brunnera macrophylla, 226
buckthorn, sea
SEE *Hippophae rhamnoides*
Buddleja davidii, 120
bugle SEE *Ajuga reptans*
bulb-planting tools, 93
bulbs and corms, 19–21, 156–79
autumn, 106–7
planting, 93–4
Bupleurum rotundifolium, **50**
'burn-off' method, clearing ground, 101, 105
burnet moths, 123, 126
burnet-saxifrage
SEE *Pimpinella saxifraga*
burning bush SEE *Euonymus alatus*
burning meadows and prairies, 55, 66, 105, 193, 194, 199, 201
Burrell, C. Colston, 196, 198–9, 202, 216, 228, 240
buttercups SEE *Ranunculus*
butterflies, 10, 14, 16,**16**, 24, 38, 49, **64**, 66, 69, 77, 108, 115, 120, 123, 124, 199, 211

C

calcium, 106, 202
Calendula officinalis (pot marigold), **188**, 189
Calke Abbey, Derbyshire, 93
Calluna vulgaris (ling, heather), 82
Caltha palustris (kingcup, marsh marigold), 83, 133, 204
C.p. var. *palustris*, 133, **133**
camas, death SEE *Zigadenus*
Camassia (quamash), 73, 80, 175, 204–5
C. angusta (prairie hyacinth), 175, 205
C. cusickii, **174–5**, 175
C. leichtlinii, 175
C.l. subsp. *suksdorfii* 'Electra', 175
C. quamash, **1**, **73–4**, 175
C. scilloides (wild hyacinth), 175, 205
Cambridge, 155
Cambridge Botanic Garden, 172
Cambridgeshire, 173
Campanula (bellflower), 155
C. lactiflora, 155
C. rapunculoides (creeping bellflower), 155
C. rotundifolia (harebell), 25, 155, 173
campion SEE *Silene*
canna, 224
Cardamine pratensis (lady's smock, cuckooflower), 17, 83, **83**, 133, **144**, 145
C. raphanifolia (greater cuckooflower), 145
Carduus nutans (musk thistle), 70
Carex (sedges), 80, 119, 195
C. elata 'Aurea' (Bowles's golden sedge), **82**
C. vulpina (foxtail sedge), 69, 131
Carlina vulgaris (carline thistle), 69, 126
carnations SEE *Dianthus caryophyllus*
carrot, wild SEE *Daucus carota*
Carson, Rachel, *Silent Spring*, 28
Carson Pass, Nevada, 175
catch-crops, annuals, 95
catchfly SEE *Silene*
Cato, 58
cat's ear SEE *Hypochaeris radicata*
celandine SEE *Ficaria verna*
Centaurea cyanus (cornflower, bachelor's buttons), **50–51**, 51, **63**, **181**, **184**, 184–5, 186, 189
C. nigra (lesser knapweed), **8–9**, 22, 42, **42**, 59, 77, 83, 97, 127, **197**,
C. sadleriana, 70, 129
C. scabiosa (greater knapweed), 70, 97, 127
Centaurium erythraea (common centaury), 97, 152
Centranthus ruber (red valerian), **78, 86**, 120
Cerastium fontanum (common mouse ear), **215**
chalk downland, 56, **56**, 66, **101–2**, 113, 126
Chamerion angustifolium 'Album' (white rosebay willow herb), 105
chamomile, corn SEE *Anthemis arvensis*
Charles, Prince of Wales, 39, 170
Chatto, Beth, 135, 178, 196, 203
Chicago, 204
chickweed, 187
Cichorium intybus (chicory), 220
cinquefoil, creeping SEE *Potentilla reptans*
Cirsium (thistles), 11, 55, 89, 99, 186, 217
C. arvense (creeping thistle), 106, 126, 176, 220
C. dissectum (meadow thistle), 127
C. vulgare (spear thistle), 127
Clare, John, 122
clary, wild SEE *Salvia verbenaca*
clay subsoil, 95, 96
Clematis viticella, 11
climate, 17, 49, 55–6, 89, 101, 146, 161, 170, 177, 178, 192, 203, 206, 209, 212
climax vegetation, 55
Clinopodium vulgare (wild basil), 153
clover
prairie SEE *Dalea purpurea*
red SEE *Trifolium pratense*
sweet SEE *Melilotus alba*
white SEE *Trifolium repens*
Cluny museum, Paris, 57
cock's foot SEE *Dactylis glomerata*
Coeloglossum viride (frog orchid), 140
Colchicum (meadow saffron), 70, 106, 107, 108, 161

C. agrippinum, 161
C. autumnale, 70
Colonna, Francesco, *Hypnerotomachia Poliphili*, 59
Colorado, 203
Columbia Gorge, Portland, Oregon, **223**
columbine SEE *Aquilegia*
comfrey SEE *Symphytum officinale*
common land, 57
compass plant SEE *Silphium laciniatum*
composites, 126–31, 206–12
compost, 98, 105, 108, 110,
potting, 94, 145
concrete, crushed, 96
coneflower SEE *Rudbeckia hirta*
greyhead SEE *Ratibida pinnata*
pale SEE *Echinacea pallida*
prairie SEE *Ratibida columnifera*
purple SEE *Echinacea purpurea*
Conopodium majus (pignut), 151
conservation, 14, 33, 35–43, 49, 70, 73, 74, 136, 140, 215, 216–17
Consolida (larkspur)
C. orientalis (eastern larkspur), 191, **191**
C. regalis (branched larkspur), 191
Common, Bob, 30
common blue butterfly (*Polyommatus icarus*), **16**
convolvulus, field, 95
copper butterflies, small, **64**, 66
coppice woodland, 15, 127, 141, 174, 225, 226
cord grass, prairie SEE *Spartina pectinata*
Coreopsis grandiflora (tickseed), 210
C. lanceolata, 201
C. palmata, 210
C. tinctoria (annual tickseed), 190, **220**, 221
C. tripteris (tall tickseed), **200**, 201
corms SEE bulbs and corms
corn chamomile SEE *Anthemis arvensis*
corn cockle SEE *Agrostemma githago*
corn marigold SEE *Glebionis segetum*
cornfield plants, 45, **50**, 62, 180–91, **181**
cornflower SEE *Centaurea cyanus*
Cornus alba (dogwood), 135, 167
Cornwall, 120, 135
Coronilla varia (crown vetch), 69, 124
Coronation Meadows project, 39
Cortaderia selloana 'Pumila' (pampas grass), 24, 199, 211
Cosmos, 224
C. bipinnatus, 187, 190, 191
Cotinus (smoke bush), 226
C. coggygria, **85–86**
couch grass SEE *Elymus repens*
cow parsley SEE *Anthriscus sylvestris*
cow wheat SEE *Melampyrum arvense*
cowslip SEE *Primula veris*
coyotes, 151
Coyte, Dr W.B., 60
cranesbills SEE *Geranium*
crab apple trees, **12**
Crataegus (hawthorn), **12–13**, 147, 178, 131
C. orientalis, 161
Crepis (hawksbeard), 129
crested dog's tail SEE *Cynosurus cristatus*
crickets, 14, 16, 34, **108**
Crocus, 11, 74, 79, **80**, 106, 108, **156–7**, 158–61
C. chrysanthus, 19, 158, 160
C.c. 'Snow Bunting', 158, **159**
C. flavus subsp. *flavus*, 158
C. imperati, 160
C. kotschyanus, 161
C. nudiflorus, 107, 160, 161
C. sieberi 'Violet Queen', 160
C. speciosus, **160**, 161
C. tommasinianus, 19, 114, **114** 158
C.t. 'Whitewell Purple', 158
C. vernus (Dutch crocus), 19, **19**, 79, 80, 158
crowsfoot, 59
crucifers, 145
cuckooflower SEE *Cardamine pratensis*
culver's root SEE *Veronicastrum virginicum*
cup plant SEE *Silphium perfoliatum*
Curtis, William, *Flora Londonensis*, 60
Curtis Prairie, Madison, Wisconsin, 195–6
Cuscuta (dodder), **194**
cutting SEE mowing
Cyclamen hederifolium, 178
Cynosurus cristatus (crested dog's tail), 117
Cyrus the Great, King of Persia, 60

D

Dactylis glomerata (cock's foot), 114
Dactylorhiza fuchsii (spotted orchid), 24, **36–37**, **52–3**, 81, **136**, 140, **140**, 141
D. × *grandis*, 96
D. praetermissa (southern marsh orchid), 141
daffodils SEE *Narcissus*
daisies, 25, 47, 55, 60, 80, 126–31, 206–11
moon SEE *Leucanthemum vulgare*
ox-eye SEE *Leucanthemum vulgare*
Tahoka SEE *Machaeranthera tanacetifolius*
Dalea candida (white prairie clover), 206
D. purpurea (violet prairie clover), 206
La Dame à la Licorne tapestry, 57
dandelions SEE *Taraxacum officinale*
Daphne pontica, 11
Darmera peltata (water saxifrage), 146, **146**
Datchler, Keith, 38–9
Daucus carota (wild carrot), 152, 188
death camas SEE *Zigadenus*
deer, 55, 196
Delphinium carolinianum (prairie larkspur, Carolina larkspur), 195
Derbyshire, **114**, 135, 142, **215**
Deschampsia cespitosa, 70
devil's-bit scabious SEE *Succisa pratensis*
Dianthus (pinks), 142–3
D. armeria (Deptford pink), 143
D. carthusianorum, 25, 70, 142–3
D. caryophyllus (carnation), **54**
D. deltoides (maiden pink), 143
D. giganteiformis subsp. *pontederae*, 70, 143
Dibbol, Neil, 198, 199
Dichelostemma congestum, **223**
Digitalis (foxglove), 154, 225
D. ferruginea, 226
D. purpurea, 41, 221
disturbed ground, 66, 70, 99, 121, 126, 145, 152, 218, 221
Dixter SEE Great Dixter
dock SEE *Rumex*
prairie SEE *Silphium terebinthinaceum*
dodder SEE *Cuscuta*
Dodecatheon meadia (shooting star), 204
D.m. f. *album*, 204
dog-violet, common SEE *Viola riviniana*
dog's-tooth violet SEE *Erythronium dens-canis*
dogwoods SEE *Cornus alba*
Doronicum (leopard's bane), 92
D. orientale, 226
D. pardalianches, 131
downlands, 56, **56**, 66, 101–2, 113, 126,
dragonflies, **55**
drainage, 17, 82–3, 88, 94, 142, 148, 163, 202

dropseed, prairie SEE *Sporobolus heterolepis*
dropwort SEE *Filipendula vulgaris*
water SEE *Oenanthe crocata*
drought, 49, 55, 65, 94, 121, 126, 127, 161, 182, 194, 202, 203,
dry prairies, 194, **194**, 195, 202, 203, 204, 205, 206, 210
Dungeness, Kent, 41, 145
Dunnett, Dr Nigel, 45, 89, 186, 187, 189, 190, 191, 220, 221, 225, 226, 227
Dutch crocus SEE *Crocus vernus*
Dyer, Geoffrey, 47
dyer's greenweed SEE *Genista tinctoria*

E

Echinacea (coneflower), 194, **208–9**, 165–6
E. angustifolia, 210
E. pallida (pale coneflower), 195, 201, 210
E. purpurea (purple coneflower), **200**, 209, **210**
Echium vulgare (viper's bugloss), **185**, 189
edgings, 86, 198
elder SEE *Sambucus*
elk, 194, 196
Elymus repens (couch grass, quack, twitch grass), 94, 115, 131, 186, 198
Emorsgate Seeds, 97, 100
English Channel, 119, 138
English iris SEE *Iris latifolia*
Epilobium (willowherbs), 149–50
E. angustifolium (rosebay willowherb), 149–50
Eranthis hiemalis (winter aconite), 135
Erodium manescavii (storksbill), 150
Eryngium maritimum (sea holly), 18
E. yuccifolium (rattlesnake master), 24, 199, **212**, 212–3
Erysimum linifolium, **83**
Erythronium dens-canis (dog's-tooth violet), 19, 20, **20**, 21, 83, 92, **92**, 165, 171, **171**,
E. revolutum (American trout lily), 83, **93**, 171
escaped plants, 89 **89**, 119–10, 187, 210
Eschscholzia californica (California poppy), 189, **223**
Euonymus alatus (burning bush), **210**
Eupatorium maculatum, 119, 204
E.m. Atropurpureum Group, 24, 211
E.purpureum (Joe Pye weed), 119, 211
E.p. 'Album', 211
Euphorbia (spurge)
E. corollata, 213
E. palustris, 101
E. paralias (sea spurge), 18
E. x *pseudovirgata, 22, 23*
E. seguieriana, **67**, 69
Euphrasia (eye-bright), 97, 154
E. officinalis, 154
evergreens, 85
eye-bright SEE *Euphrasia*

F

false aster SEE *Boltonia asteroides*
false indigo, blue SEE *Baptisia australis*
false oat-grass SEE *Arrhenatherum elatius*
farming, 35, 38, 39, 41, 55, 65, 66, 215
felwort SEE *Gentianella amarella*
Ferguson, Amanda, 14, 65
ferns, 11, 141–2
fertile soil, creating a permanent meadow, 95–6
Ferula communis (giant fennel) 21
fertilizers, 62, 65, 85, 92, 142, 182, 215, 217
Festuca (fescues), 22, 29, **64**, **65**, 115
F. arundinacea (tall fescue), 69, 114, 115, 131
F. idahoensis, **13**
F. pratensis (meadow fescue), 115
F. pseudovina (dwarf fescue), 65, 115
F. rubra (red fescue), 115
F. rubra subsp. *commutatus* (Chewing's fescue), 115
F.r. subsp. *litoralis* (creeping red fescue), 115
Ficaria verna (celandine), 79, **79**, 132
F.v. Flore Pleno Group, **132**
figwort SEE *Scrophularia nodosa*
Filipendula, 119
F. purpurea, 148
F. rubra 'Venusta' (queen of the prairie), 83, 148
F. ulmaria (meadowsweet), 17, 79, 148, **148**
F. vulgaris (dropwort), 66, 148
fire, 95, 194–5, 196
firewheel SEE *Gaillardia pulchella*
flag iris, yellow SEE *Iris pseudacorus*
flax, red SEE *Linum grandiflorum* 'Rubrum'
Florence, 58
flowery medes, 57–60
Flymo, 111
forbs, prairies, 193, 194, 196, 198, 201, 204, 227
foreign species, 114, 119–20
foxglove SEE *Digitalis purpurea*
foxtail, meadow SEE *Alopecurus pratensis*
foxtail grass SEE *Setaria macrochaeta*
France, 138, 216
Frenich, Perthshire, 177
Fritillaria, 167
F. meleagris (snakeshead fritillary), **2**, **58**, 59, **78**, **79**, **120**, 132, 148, 162, 167, **168–9**
F.m. var. *unicolor* subvar. *alba*, **167**
fritillary, snakeshead SEE *Fritillaria meleagris*
frogs, **55**

G

Gaillardia pulchella (Indian blanket, firewheel), 190–1
Galanthus (snowdrop), 11, 74, 92, 108, 120, 157–8
G. nivalis, 157
G.n. 'S. Arnott', 157
Galatella linosyris, 66
G. sedifolius, 66
Galega officinalis (goat's rue), 206
G.o. 'Alba', 199
G. orientalis, 22
Galium (bedstraws), 154–5
G. mollugo (hedge bedstraw), **218**
G. palustre (marsh bedstraw), 70
G. saxatile (heath bedstraw), 154
G. verum (lady's bedstraw), 97, 99, 154–5,
Gammons, Nikki, 41, 42
Garden House, Devon, 152
garlic SEE *Allium*
honey SEE *Nectaroscordum siculum*
Garrett, Fergus, 65, 73, 74, 94, 100,106, **109**, 110, 132, 145, 158
Aysegul, **31**
gayfeather, spike SEE *Liatris spicata*
Genista tinctoria (dyer's greenweed), 24, **56**, 125
gentian, autumn SEE *Gentianella amarella*
Gentiana, 152
G. acaulis, 152

Gentianella amarella (autumn gentian, felwort), 152
Georgia, 224
Geotextile, 222
Geranium (cranesbills), 11, 150
G. clarkei 'Kashmir White', 150
G. endressii, 150
G. himalayense, 150
G. maculatum, 159
G. pratense (meadow cranesbill), 30, 124, 150, **218**
G. psilostemon, 101, 150, 182
G. sylvaticum (wood-cranesbill), 150, 226
Gerard, John, 167
Germany, 143
germination, 39, 94, 98, 186, 196, 199, 201, 217
Gladiolus communis subsp. *byzantinus*, 77, **78, 81**, 177–8, **178–9**
Glebionis segetum (corn marigold), 57, **180–1**, 181, 184, 184, 185, 187, 189, 190
G. coronariua, 189
G.c. 'Primrose Gem', 189
globeflower SEE *Trollius*
glyphosate, 44, 93, 94, 186, 199, 217
goat's rue SEE *Galega officinalis*
goldenrod SEE *Solidago*
goldilocks SEE *Ranunculus auricomus*
gorse, 82
grape hyacinth SEE *Muscari armeniacum*
grasses, 66–70
aggressive grasses, 114–5, 227
annual grasses, 191
choosing seeds, 96–7
ecological restoration, 216
mowings, 91, 111, 121
non-competitive grasses, 115–7
prairie, 199, 202–4
reducing vigour, 29–30 86, 97, 113, 125, 154, 217
grasshoppers, 14, 16, 34, 49, 108
Gravetye Manor, Sussex, 61
Gray, Alan, 181, 185
grazing, 32–3, 33–5, 55, 56–7, 66, 105, **106**, 194, 196, 217, 220
Great Dixter, East Sussex 7–47, **9, 42**, 50–51, 72–83, 85–6, 106, **107**, 113, 119, 120, 125, 126, 130, 132–3, 138, 140, 141, 142, 145, 150, 153, 157, 160, 170, **170**, 172, 174, 177, 190, 224, 225
Botticelli garden, 58, **58**, 74, 78, 79, 136, 139, 167, 173
front meadows, **1**, **7**, **8–9**, 79–80, **80**, 86, 108, **109**, 113, **157**,
the hollow, 83, **83**
horse pond, 17–18, **18**, 19, 20, 21, 82–3, **82**, 111, 135, 143, 148, 155, **155**, 167, 177,
orchard meadows, **2**, **4**, **11–12**, 19, 29, 30, **47**, **51**, **52–3**, 74–9, **76–7**, **86**, **87**, **88**, 89, 90, 93, 107, 108, **111**, 157, 158, 160, 162,**162**, 163, 167, 174, 178,
topiary lawn, 16, 24–8, **25**, **26–7**, **28**, 80–81, **81**, **85**, 89, 108, 116, 117, 122, 129, 140, 141, 155,
upper moat, 17, **58**, **78**, 79, **81**, 107, 134, 139, 146, 148, 161, 166, 174, 177,
Great Dixter Charitable Trust, The, 47
groundsel, 187
gunnera, 111
Gymnadenia conopsea (fragrant orchid), **56**, 138
Gypsophila elegans (baby's breath), 189–90
G. paniculata, 69

H

hairgrass, tufted, SEE *Deschampsia cespitosa*
Hardwick Hall, Derbyshire, 142
harebell SEE *Campanula rotundifolia*
Harlow Carr, Yorkshire, 101, 106, 225, **227**
Harvey, Graham, 24, 80
Hassam, Childe, 182
Hastings, East Sussex, 222
hawkbit SEE *Leontodon*
hawksbeards SEE *Crepis*
SEE *Hieracium* orange
SEE *Pilosella aurantiaca*
hawthorn SEE *Crataegus*
hay, 16, 30, **31**, 32, 33, 35, 38, 39, 49, 55, 57, 66, 75, 96, 96, 98, 100, **108**, 114, 121, 123, 130, 195, 215, 217
Hay Meadows Report, 121
hay rattle SEE *Rhinanthus minor*
Haydock, Catherine, 42
hay strewing, 30, **31**, 96, 100
heather SEE *Calluna vulgaris*
Helenium, 194
H. autumnale (sneezeweed), 206
H. flexuosum (purple-headed sneezeweed), 206
Helianthus (sunflowers), 187, 196, 204, 206, 211
H. annuus, 191
H.a. 'Valentine', 191
H. debilis, 191
H.d. 'Vanilla Ice', 191
H. grosseserratus (sawtooth sunflower), 24, **197**, 199, 206
H. maximiliani, 206
H. occidentalis (western sunflower), 206
H. salicifolius (willow-leaved sunflower), 206
H. tuberosus (Jerusalem artichoke), 196, 206
hellebores, 11, 92
helleborines, 141
Heracleum (hogweed), 88, 141
H. mantegazzianum (giant hogweed), 152
H. sphondylium, 152
Herat (Afghanistan), 60
herbicides, 51, 62, 65, 126, 129, 144, 145, 146, 150, 154, 182, 183, 186, 191, 196, 198, 199, 201, 223, 224
Hesperis matronalis (sweet rocket), 221, 226
Hieracium (hawkweed), 129
Highgrove, Gloucestershire, 170
highway verges, 47, 124, 146, 222, **222**, 224
Himantoglossum hireinum (lizard orchid), 138
Hippocrepis (horseshoe vetch), **56**
Hippophae rhamnoides (sea buckthorn), 226
Hitchmough, Dr James, 56, 60–1, 74, 155–7, 157, 163, 164, 166, 167
Hobhouse, Penelope, 60
hogweed SEE *Heracleum*
Holcus lanatus (Yorkshire fog), 97, 114, **114**, **219**
H. mollis (creeping soft grass), 115
Holme Pierrepoint Hall, Nottingham, **170**
honesty SEE *Lunaria annua*
honey garlic SEE *Nectaroscordum siculum*
Horace, 58
Hordeum (barley), 181, 184
Horsemint SEE *Monarda*
horseshoe vetch SEE *Hippocrepis*
Hortobágy National Park, Hungary, **64–5**, 65–6
Hudson, W.H., 168
Hungary, **62**, **64–5**, 65–71, **67**, **68–9**, 89, 119, 124, 126, 129, 131, 153, 172, **185**,191, **191, 222**

The Hunt of the Unicorn as Lover tapestry, **55**, 57
hyacinth SEE *Camassia*
Hyacinthoides (bluebell), 57, 83, 120, 173–4, 225
  H. hispanica (Spanish bluebell), 139, 174
  H. non-scripta, 173, **173**, 174, 181
Hypochaeris radicata (cat's ear), 100, 129, **215**

I
Ibn Arabshah, 60
Ice Age, 55, 119
Iceland, 216
India, 60
Indian blanket SEE *Gaillardia pulchella*
Indian grass SEE *Sorghastrum avenaceum*
Indian paintbrush SEE *Castilleja miniata*
insects, 14, 15, 17, 18, 33, 34, 35, 86, 105, 106, 127, 215, 217,
Inula magnifica, 29, **76–7**, 79, 131, **131**
invertebrates, 35, 77, 108, **108**
Iran, 60
Ireland, 135
Iris, **55**, 89, 177
  I. latifolia (English iris), 79, 177
  I. pseudacorus (yellow flag iris), **82**, 83, 151, 177,
  I. sibirica, **82**, 177
  I. virginica, 195
    I.v. var. *shrevei*, 177
Iron Age, 56
ironweed SEE *Vernonia*
Ismelia carinata, 189
Italy, 58, 167

J
Jacobaea vulgaris (ragwort), 55, 105–6, 130
Jäger, Hermann, *Lehrbuch der Gartenkunst*, 62
Jensen, Jens, 195
Jewell, David, 20
Jerusalem artichoke SEE *Helianthus tuberosus*
Joe Pye weed SEE *Eupatorium purpureum*
Johnson, Lady Bird, 190, 223
Juncus (rushes), 83, 143

K
Kay, Quentin, 100
Keats, John, 129
Keen, Mary, 135
Ketley's, Sussex, **112**
Kew Gardens, Surrey, 93, 141, 172
King, Jane, 101, 106
kingcups SEE *Caltha palustris*
Kingston Maurward, Dorchester, Dorset, 178
knapweed SEE *Centaurea*
Knautia arvensis (field scabious), 97, 125, **125**
Knowsley, Merseyside, 95, 97, 99, 106, 114
Kosa, Geza, 65

L
labiates, 153
Lady Bird Johnson Wildflower Center, Austin, Texas, 190
Lady Farm, Somerset, **99**
lady's smock SEE *Cardamine pratensis*
Landlife, Liverpool, 95, 111, 218–20
landscape designers, 45, 225
larkspur SEE *Consolida*
  Carolina SEE *Delphinium carolinianum*
Lathraea clandestina (toothwort), 155, **155**
Lathyrus nissolia (grass vetchling), 121
L. pratensis (meadow vetchling), 100, 120–1
  L. tuberosus (tuberous pea), 66, 121, **121**
Lavatera trimestris (rose mallow), 191
lawns, 28, 45, 49, 50, **73**, 88, 121, 136, 129, 133, 144, 153, 154
lead plant SEE *Amorpha canescens*
leek, round-headed SEE *Allium sphaerocephalon*
legumes, 69, 120–125, **123**, 205–206
Leontodon (hawkbit), 129
  L. autumnalis (autumn hawkbit), 77, 97, **128**, 129
  L. hispidus (rough hawkbit), 100, 129, **215**
leopard's bane SEE *Doronicum*
Leucanthemum vulgare (ox-eye daisy, moon daisy), 8, 16, 24, 25, **25**, 28, 32, 39, 57, 89, **98–9**, **101–2**, **125**, 126, 216, 220, 221, **221**, **224**
Leucojum aestivum (summer snowflake), **78**, 166–167, **166**
  L.a. subsp. *aestivum* (Lodden lily), 167
  L.a. 'Gravetye Giant', **166**, 167
  L.a. subsp. *pulchellum*, 167
Lewis, Pam, 95, 96, 106
Liatris, 198–9
  L. punctata (dotted blazing star, gayfeather), 211
  L. pycnostachya (prairie blazing star), 201, 211, 167
  L. spicata (dense blazing star, spike gayfeather), 211, **211**
Liddell, Molly, 163
lilies
  American trout SEE *Erythronium revolutum*
  Lent SEE *Narcissus pseudonarcissus*
  Lodden SEE *Leucojum aestivum* subsp. *aestivum*
  Madonna SEE *Lilium candidum*
  Michigan SEE *Lilium michiganense*
  prairie SEE *Lilium philadelphicum*
  triplet SEE *Triteleia laxa*
  Turk's cap SEE *Lilium michiganense*
Lilium, 177
  L. candidum (Madonna lily), **55**
  L. martagon, 177
  L. michiganense (Michigan lily, Turk's cap lily), 205
  L. philadelphicum (prairie lily), 205
  L. pyrenaicum, **176**, 177
  L. superbum, 205
Limnanthes douglasii (poached egg flower), **89**
Limonium gmelinii (sea lavender), 66
Linaria maroccana (toadflax), 189, 221
ling SEE *Calluna vulgaris*
Linum grandiflorum 'Rubrum' (red flax), **188**, 189
Listera ovata (twayblade orchid), **136**, 140
Liverpool, 218
Lloyd, Christopher (Christo), 7–14, 16–22, 24–5, 28, 29–30, 45, 47
Lloyd, Daisy (Christo's mother), 58, 59, 61, 73, 74, 75, 77, 79, 80, 85, 127, 136, 138, 139, 140, 146, 150, 163, 166, 167, 171, 173
Lloyd, Nathaniel (Christo's father), 19, 61, 74, 79, 80, 81, 82, 85, 116
local plants, using seed from, 216
Lolium perenne (perennial rye grass), 22, 29, 88, 114, 222
Lomatium martindalei, **223**
loosestrife
  purple SEE *Lythrum salicaria*
  yellow SEE *Lysimachia vulgaris*
Lord, Tony, 86, 88

Lotus corniculatus (bird's-foot trefoil), 8, **16**, 22, 24, 28, 42, 59, 97, 122, **123**,
lousewort SEE *Pedicularis sylvatica*
love-in-a-mist SEE *Nigella arvensis*
Lunaria annua (honesty, money flower), 226
Lupinus (lupin or lupine), 124–5, **124**, 220
L. perennis, 205
L. polyphyllus, 124
L. x *regalis* (Russell lupins), 124
L. subcarnosus (Texas bluebonnets), 190, 224
L. succulentus, 190
Lutyens, Edwin, 24, 43, 85, 86, 89
Lychnis flos-cuculi (ragged Robin), 143, **143**
L. oculata, 190
Lydden Down, Kent, **56**
Lysimachia vulgaris (yellow loosestrife), 70
Lythrum salicaria (purple loosestrife), 119–20

M

Mabey, Richard, 135
Machaeranthera tanacetifolia (Tahoka daisy), 190
machinery, mowing, 60, 108, 110, **110**
MacIntyre, Donald, 100
Magdalen College, Oxford, 167, **168–9**
mallow, rose SEE *Lavatera trimestris*
marigolds
corn SEE *Chrysanthemum segetum*
marsh SEE *Caltha palustris*
pot SEE *Calendula officinalis*
marram grass SEE *Ammophilia arenaria*
Mary, Virgin, 57, 58
meadows
annual cycle, 217
annual meadows, 44, 180–91
at Dixter, 72–83
bulbs and corms, 156–179
creating a permanent meadow, 90–103
edging, 40–4, 86, 198
management, 104–111
paths, **4**, 7, **7**, 8, **73**, 74, 79, 80, **81**, 86, **87**, 88, **88**, 113, **157**, 170, 175
perennials, 118–155
SEE ALSO mowing
meadowsweet SEE *Filipendula ulmaria*
medes, flowery, 57–60
Medicago lupulina (black medick), 16
medieval gardens, 57–60
Melampyrum arvense (cow wheat), 66, 70, **116, 125**
M. barbatum, 69
Melilotus, 59
M. alba (sweet clover), 196
Mendips, Avon/Somerset, 56
Mentha requienii (Corsican mint), 153
Merseyside, 95, 106
mesic prairies, 172, 195, 203, 205, 206
metaldehyde, 201
Metropolitan Museum of Art, New York, 57
Michaelmas daisies, 203, 210
Michigan, Lake, 172
mignonette SEE *Reseda*
milkweed SEE *Asclepias*
milkwort, chalk SEE *Polygala calcarea*
millet, purple SEE *Panicum miliaceum* 'Violaceum'
Minnesota, 74, 193, 196, 198, 203, 206, 216
Minnesota Landscape Arboretum, 196, 198
mint, Corsican SEE *Mentha requienii*
Miscanthus, 211
Molinia hungarica, 70
Monarda (bee balm, Oswego tea, bergamot, horsemint), 194, 196, 213
M. didyma, 213
M. fistulosa, 196, **200**, 201, 204, 213
Monet, Claude, 182
money flower SEE *Lunaria annua*
monoculture, 38, 224
moon daisy SEE *Leucanthemum vulgare*
moonwort SEE *Botrychium lunaria*
Morrison, Pip, 141
mosses, 10, 35, 142
mouse ear SEE *Cerastium fontanum*
mowing
creating a permanent meadow, 91, 99
established meadows, 105, 106–111
machinery, 60, 108, 110, **110**
managing a sown area, 99
mechanics of cutting, , 108–9, **109**
paths, 86, **87**
picking up mowings, 119–11
prairies, 201
when to cut, 106–8
Muscari (grape hyacinth)
M. armeniacum, 59, 173
M. neglectum, 173

N

Narcissus (daffodils), 11, **12–13**, 57, 58, **61**, 62, 75, **75**, 77, **90**, 108, 158, 162–6, 170, 223
N. asturiensis, 165, **165**
N. bulbocodium (hoop-petticoat daffodil), 108, 129, 165
N.b. var. *citrinus*, 165, **165**
*N.b.*subsp. *bulbocodium* var. *conspicuus*, 20, **21**, 162
N. cyclamineus, 20, 21, 166
N. 'Emperor', **75**, 162, **163**
N. 'Lillie Langtry', 162
N. 'Minnie Hume', 162
N. minor, 20, 83, **92**, 163–5, **165**, 171, **171**
N. poeticus var. *recurvus* (poet's narcissus, pheasant's eye narcissus), 74, **76–7**, 162, **165**
N. 'Princeps', 74, 162
N. pseudonarcissus (Lent lily), 80, **80, 156–7**, 163, **165**, 167
National Wildflower Research Center, Austin, Texas, 190
native plants, 62, 98, 100, 101, 119–20, 187, **194**, 195, 196, **209**, 216, 222, 223–4
naturalizing plants, 20, 118
Nebraska, 216
Nectaroscordum siculum (honey garlic), 172
Nemophila menziesii (baby blue eyes), 190
Neolithic man, 56
Netherlands, 33, 57
nettles SEE *Urtica dioica*
Newborough Warren, Anglesey, 18
New York Botanic Garden, 201, 203
New Zealand, 124, 216
Nicandra physalodes (shoo fly plant), 144
Nigella (love-in-a-mist), 89
N. arvensis, 58
N. damascena 'Oxford Blue', 189
nitrogen, 35, 62, 79, 91, 94, 106, 120, 182, 205
none-so-pretty SEE *Silene armeria*
Nordic Gene Bank, 216
Normans, 57
North America, 1, 55, 66, 89, 114, 119, 120, 175, 185, 190, 191, 206, 216, 223, 226
SEE ALSO prairies; United States of America
Northiam, Sussex, 7, 47, 61
Northiam Horticultural Society, 30

O

oak trees, 7, 11, 20, 21, 82, 135, 165, 195
oats SEE *Avena*
obedient plant SEE *Physostegia virginiana*
Odontites vernus (red bartsia), 97
Oenanthe crocata (water dropwort), 17, 151, **151**
Old Vicarage, East Ruston, Norfolk, **180–1**, 185
onions SEE *Allium*
Oostvardersplassen project, Netherlands, 33
Ophioglossum vulgatum (adder's tongue fern), 100, 141–2, **141**
Ophrys apifera (bee orchid), 59, 96, 97, 101, **118**, 138
orach, purple SEE *Atriplex hortensis*
orange tip butterfly (*Anthocharis cardamines*), 17, 83, 144
orchards, 50, 60, 74, 162
orchids, 11, 16, 22, **25**, 28, 32, 39, 77, 81, 97, 108, 136–41, 162
  bee SEE *Orphrys apifera*
  common spotted SEE *Dactylorhiza fuchsii*
  early-purple SEE *Orchis mascula*
  fragrant SEE *Gymnadenia conopsea*
  frog SEE *Coeloglossum viride*
  greater butterfly SEE *Platanthera chlorantha*
  green-winged SEE *Anacamptis morio*
  lizard SEE *Himantoglossum hircinum*
  pyramidal SEE *Anacamptis pyramidalis*
  southern marsh SEE *Dactylorhiza praetermissa*
  twayblade SEE *Listera ovata*
  western prairie fringed SEE *Platanthera praeclara*
Orchis, 59, 138
  O. mascula (early-purple orchid, male orchid), **54**, 59, 135, **136**, 138–9, 174, 225
Orlaya grandiflora, **185**, 189
Ornithogalum (star-of-Bethlehem)
  O. nutans, 172–3
  O. umbellatum, 173
Oryzopsis miliacea, 11
Oswego tea SEE *Monarda*
Our Lord's candle SEE *Yucca*
ox-eye daisy SEE *Leucanthemum vulgare*

P

pampas grass SEE *Cortaderia selloana* 'Pumila'
panic grass SEE *Panicum virgatum*
Panicum, 203
  P. miliaceum 'Violaceum' (purple millet), 190, 191
  P. virgatum (switch grass, panic grass), 195, 196, 204, 216
    P.v. 'Hänse Herms', 203
    P.v. 'Prairie Sky', 204
    P.v. 'Rubrum', **203**
    P.v. 'Shenandoah', 204
Papaver (poppies), 11, 51, 60, **63**, 181, 189, **220**
  P. commutatum (ladybird poppy), 190
  P. dubium, 190
  P. orientalis (oriental poppy), 21, 22
  P. rhoeas (common or field poppy), **51–2**, **180–1**, 184, 185, **185**, 187, 190,
    P.r. Shirley Group, 190
  P. somniferum (opium poppy), 190
parsnip, wild SEE *Pastinaca sativa*
Pasargadae, Iran, 60
pasqueflower SEE *Pulsatilla*
pasture, 10, 14, 20–1
paths, **4**, 7, **7**, 8, **73**, 74, 79, 80, **81**, 86, **87**, 88, **88**, 113, **157**, 170, 175
Pastinaca sativa (wild parsnip), 151
pea, tuberous SEE *Lathyrus tuberosus*
Peak District, **114**, 117, 119, 121, 123, **215**
pearlwort SEE *Sagina procumbens*
Pear trees, **12**
Pedicularis sylvatica (lousewort), 154
Penner, Henry, 222
Pentridge House, Dorset, **102–3**, **106**
perennial weeds, 183, 186–7, 209, 217, 220
perennials, 118–155
Persia, 60
Persicaria bistorta 'Superba' (bistort), 101
Peucedanum officinale (sulphurwort), 66
Phacelia campanularia (California bluebell), 190
  P. tanacetifolia (lacy scorpion weed), 184
pheasants, 132, 167
pheasant's eye narcissus SEE *Narcissus poeticus* var. *recurvus*
Phlomis tuberosa, 69
Phlox pilosa (prairie phlox), 204
phosphorus, 106, 182
Phragmites australis (common reed), 30, 70
Physostegia virginiana (obedient plant), 153
Pickering's Pasture, Liverpool, 95, 218–20
pignut SEE *Conopodium majus*
Pilosella aurantiaca (orange hawkweed), 130, **130**
Pimpinella saxifraga (burnet-saxifrage), 151
pinks SEE *Dianthus*
Place House, Peasmarsh, 163
Plantago (plantain)
  P. lanceolata (ribwort plantain), 32, **32**, 153, **154**
  P. major (greater plantain), 153
  P. media (hoary plantain), **49**, 153
plantain SEE *Plantago*
Platanthera chlorantha (greater butterfly orchid), 141
  P. praeclara (western prairie fringed orchid), 138
Pliny the Younger, 58
ploughing, annual meadows, 183
Poa (bluegrass), 196, 198
poached egg flower SEE *Limnanthes douglasii*
poet's narcissus SEE *Narcissus poeticus* var. *recurvus*
pollution, roadsides, 223
polyanthus primrose, 79, 146
Polygala calcarea (chalk milk-wort), **56**
poor soil, creating a permanent meadow, 94–5
poppies
  California SEE *Eschscholzia californica*
  field SEE *Papaver rhoeas*
  ladybird SEE *Papaver commutatum*
  Mexican SEE *Argemone mexicana*
  opium SEE *Papaver somniferum*
  Shirley SEE *Papaver rhoeas* Shirley Group
porcupine grass SEE *Stipa spartea*
potassium, 106, 182
Potentilla anserina (silverweed), 147
P. erecta (tormentil), 147
P. reptans (creeping cinquefoil), 147
poverty grasses, 117
power-scythes, 108, 110
prairie blazing star SEE *Liatris pycnostachya*
prairie buttercup SEE *Ranunculus rhomboideus*
prairie clover SEE *Dalea purpurea*
prairie coneflower SEE *Ratibida columnifera*
prairie cord grass SEE *Spartina pectinata*

prairie dock SEE *Silphium terebinthinaceum*
prairie dropseed SEE *Sporobolus heterolepis*
prairie hyacinth SEE *Camassia angusta*
prairie larkspur SEE *Delphinium carolinianum*
prairie lily SEE *Lilium philadelphicum*
prairie onion SEE *Allium stellatum*
prairie phlox SEE *Phlox pilosa*
prairie shoestring SEE *Amorpha canescens*
prairies, 22–3, 192–213
  dry (xeric), 194, **194**, 195, 202, 203, 204, 205, 206, 210
  early flowers, 204–5
  establishing, 196–8
  grasses, 202–4
  maintenance, 201
  moist (mesic), 172, 195, 203, 205, 206
  North America, 194–5
  planting, 198–9
  plants, 202–13
  restoration, 195–6
  wet (hydric), 195, 196, 201, 203, 204, 205, 213
primrose SEE *Primula vulgaris*
Primula veris (cowslip), **56**, 59–60, 99,100, **102–3**, 106, 146
  P. vulgaris (primrose), 20, 59, **76**, 79, 83, 135, 146, 171, 225
  SEE ALSO polyanthus primrose
protective clothing, 110
Prunella vulgaris (selfheal), 153, **153**
public spaces, 218
Pulsatilla patens (eastern pasqueflower), 136, 204
  P. vulgaris (pasqueflower), 136

Q

quack SEE *Agropyron repens*
quaking grass SEE *Briza*
quamash SEE *Camassia*
Queen Anne's lace SEE *Anthriscus sylvestris*
queen of the prairie SEE *Filipendula rubra* 'Venusta'

R

rabbits, 55, 82
ragged Robin SEE *Lychnis floscuculi*
ragwort SEE *Jacobaea vulgaris*
rainfall, 49, 98, 101, 119, 148, 193
Ranuculus (buttercups), **1**, 8, 11, 16, 24, 25, 32, **35**, 39, 47, **76–7**, **78**, 80, 99, 106, **112**, 119, 131, 132, 135, 178, **179**
  R. acris (field buttercup, meadow buttercup), 133, 220, 226
  R. auricomus (goldilocks), 59, 83, **83**, 132–3, 135, 145
  R. bulbosus, 133
    R.b. 'Stevenii', 133
  R. fascicularis (early buttercup), 204
  R. repens (creeping buttercup), 133
  R. rhomboideus (prairie buttercup), 204
Ratibida, 199
  R. columnifera (prairie coneflower), 208
  R. pinnata (greyhead coneflower), 201, 206, **206**
rattle SEE *Rhinanthus minor*
rattlesnake master SEE *Eryngium yuccifolium*
Raven, Sarah, 189
Reader, Colin 44–5
relic grasslands, 215
Renaissance, 58–9, 60
Reseda (mignonette, weld)
  R. lutea, 70
  R. luteola, **218**
Rhinanthus minor (hay rattle, yellow rattle), 22, 24, 28, 29, **29**, 32, 35–6, 69, 80, 86, 97, **97**, 100, **102–3**, 108, 113, 154, 217
Rhine, River, 55, 167
Rhododendron ponticum, 120
Rhus (sumac)
  R. glabra, 226
    R.g. 'Laciniata', 227
  R. typhina (stag's horn sumac), 226, 227, **227**
roadsides, 24, 51, 66, 120, 125, 129, 130, 136, 147, 152, 212, 213, 215, 221–4, **221–4**, 226
Robeson, Graham, 181, 185
Robinson, William, 61–2, 74,162–3
  The English Flower Garden, 61–2, **61**, 74, 93
  The Wild Garden, 61
rocket, sweet SEE *Hesperis matronalis*
Rodrigues, Perry, 17
Romans, 56–7
Romney Marsh, Kent, 41, 56
Rosa (roses), 11, 147–8, 226
  R. 'Félicité Perpétue', 147
  R. filipes 'Kiftsgate', 147
  R. gallica, 66, **147**
  R. 'Madame Plantier', 147
  R. rugosa, 226
  R. virginiana, 226
rosinweed SEE *Silphium integrifolium*
rotary mowers, 108, 109, **109**, 111
Rothschild, Miriam, 184, 222
Roundup SEE glyphosate
Royal Botanic Gardens, Kew, Surrey, 93, 141
Royal Horticultural Society, 101, 141
Royal Society for the Protection of Birds (RSPB), 41
Rudbeckia (coneflower, black-eyed Susan), 194, 196, 206, 209, **208–9**
  R. fulgida var. *deamii*, **211**, **226–7**, 227
  R. hirta, 119, 187, 188, 190
  R. laciniata, 196
  R. subtomentosa, 201, 206
Rumex (docks, sorrels), **49**, 80, 99, 143–145, 186, 217
  R. acetosa (common sorrel), **98**, **112**, 143, 226
  R. acetosella (sheep's sorrel), 145
  R. crispus (curled dock), 145
  R. obtusifolius (broad-leaved dock), 145
rushes SEE *Juncus*
Russell lupins SEE *Lupinus* x *regalis*
rye grass, perennial SEE *Lolium perenne*

S

Sackville-West, Vita, 147
safety, using power tools, 110
saffron, meadow SEE *Colchicum*
sage, blue SEE *Salvia azurea*
Sagina procumbens (pearlwort), 142
salad burnet SEE *Sanguisorba*
Sales, John, 81, 108, 177
Salix (willow)
  S. alba var. *vitellina* 'Britzensis', **225**
    S.a. var. *sericea*, 226
  S. exigua, 226
Salvia azurea (blue sage), 195, 212
  S. nemorosa, 66, **67**
  S. pratensis, 70
  S. sclarea var. *turkestanica*, 221
  S. verbenaca (wild clary), 153
Samarkand, 60
Sambucus (elder)
  S. adnata, 148
  S. ebulus, 148
  S. nigra, 226
Sanguisorba (salad burnet), 55, **56**
  S. canadensis, 119
Saponaria officinalis (soapwort), *S.o.* 'Rosa Plena' 143
savanna, 195, 211
Savill Garden, Windsor Great Park, Surrey, 166, 171

Saxifraga (saxifrage), 59
S. granulata (meadow saxifrage), 146
saxifrage, water SEE *Darmera peltata*
Saxons, 57
Scabiosa columbaria (scabious), 126
scabious
devil's-bit SEE *Succisa pratensis*
field SEE *Knautia arvensis*
Scandinavia, 216
Scandix pecten-veneris, 59
Scènes de la Vie Seigneuriale tapestry, 57
Schizachyrium scoparium (little bluestem), 195, 203
Scilla, 11
scorpion weed, lacy SEE *Phacelia tanacetifolia*
Scotland, 82, 130, 131, 135, 146, 152, 155, 177, 225
Scott, Richard, 95, 97, 218
Scrophularia nodosa (common figwort), 154
sea buckthorn SEE *Hippophae rhamnoides*
sea holly SEE *Eryngium maritimum*
sea lavender SEE *Limonium gmelinii*
sea spurge SEE *Euphorbia paralias*
Second World War, 7, 14, 62, 82, 140
Securigera varia (crown vetch), 69, 124
sedges SEE *Carex*
seed
annual meadows, 183–184
choosing, 96–7
from hay strewing,100
prairies, 193, 198–9
sowing, 97–98
selfheal SEE *Prunella vulgaris*
semi-parasitic plants, 69, 97, 154
SEE ALSO *Rhinanthus minor*
Setaria macrochaeta (foxtail grass), 191
Seymour, Sarah, 42
Shakespeare, William, 138, 181
sheep, 34–5, 39, 57, 74, 75, 100, 106
Sheffield, 187, **209, 220**
Sheffield University, 45, 57, 101, 186, 201, 220, 225
shoo fly plant SEE *Nicandra physalodes*
shooting star SEE *Dodecatheon meadia*
Short-haired Bumblebee Project, 41
shrubs, bordering meadows, 224–6
sideoats SEE *Bouteloua curtipendula*
silage, 35, 215
Silene armeria (sweet William catchfly, none-so-pretty), 190
S. coeli-rosa (viscaria), 190
S. dioica (red campion), 120, 143, 225, 226
S. gallica (small-flowered catchfly, windmill catchfly), 190
S. latifolia (white campion), 143
Silphium, 196, 201, 208
S. integrifolium (rosinweed), **200**
S. laciniatum (compass plant), 196, 164–5
S. perfoliatum (cup plant), 24, 196, 203, 208, 164
S. terebinthinaceum (prairie dock), 24, 196, 208–9
silverweed SEE *Potentilla anserina*
Sissinghurst, Kent, 86–8, 147
slugs, 165, 201, 205
smoke bush SEE *Cotinus*
snakeshead fritillary SEE *Fritillaria meleagris*
sneezeweed SEE *Helenium*
snowdrop SEE *Galanthus nivalis*
snowflake, summer SEE *Leucojum aestivum*
soapwort SEE *Saponaria officinalis*
soft grass, creeping SEE *Holcus mollis*
Solidago (goldenrod), 119, 130, 194, 198, 203, 209
S. canadensis, 70
S. ohioensis, 201, **201**
S. rigida, 201, **201**
S. speciosa, 195, 201, **201**
S. virgaurea, 86
Sorbaria (false spiraea), 226
Sorghastrum nutans (Indian grass), 203
sorrel SEE *Rumex*
South Africa, 65, 182
South Downs, Sussex, 138
southern marsh orchid SEE *Dactylorhiza praetermissa*
Spartina pectinata (prairie cord grass), 195, 204, 206
S.p. 'Aureomarginata', 204
speedwell, slender SEE *Veronica filiformis*
spiders, 14, **15**, 34, 42, **104–5**
Spiranthes spiralis (autumn lady's tresses), **56**, 141
Sporobolus heterolepis (prairie dropseed), 203
spurges SEE *Euphorbia*
Stace, Clive, 129
Stachys officinalis (betony), 153
stag's horn sumac SEE *Rhus typhina*
star-of-Bethlehem SEE *Ornithogalum*
Sticky Wicket, Dorset, 95, **96**, 106
Stipa barbata, 69
S. borysthenica, **68–9,** 69,
S. pennata, 69
S. spartea (porcupine grass), 203
Stokes, Amanda, 209
stooling shrubs, 226, 227
storksbill SEE *Erodium manescaui*
strimmers, 106, 108, 110, 111, 187
subsoil, 44, 82, 95, 96, 199
Succisa pratensis (devil's-bit scabious), 97, 125
suckers, shrubs, 195, 226
Suffolk, 173
sulphurwort SEE *Peucedanum officinale*
sumac SEE *Rhus*
sunflowers SEE *Helianthus*
switch grass SEE *Panicum virgatum*
Symphotrichum cordifolius, 210
S. ericoides (heath aster), 210
S. laeve (smooth aster), **210**
Symphytum (comfrey), 83, **108**
S. caucasicum, 153
S. officinale, 69, 70, 131, 152–3

T

Tahoka daisy SEE *Machaeranthera tanacetifolius*
Tanacetum vulgare (tansy), 69, 131
tapestries, **55**, 57, 58
Taraxacum officinale (dandelion), 24, 43, 47, 55, 57, 129, **129**, 165
teasels, 125, 221
Texas, 190, 194, 223, 224
Texas bluebonnets SEE *Lupinus subcarnosus*
Thacker, Christopher, 57, 60
Thalictrum aquilegiifolium, 101
Thaxter, Celia, 185
thimbleweed SEE *Anemone cylindrica*
thistles
carline SEE *Carlina vulgaris*
musk SEE *Carduus nutans*
creeping SEE *Cirsium arvense*
meadow SEE *Cirsium dissectum*
spear SEE *Cirsium vulgare*
Thymus pannonicus (thyme), **64**,66
tickseed, SEE *Coreopsis*
Tifernum Tiberinum, 58
toadflax SEE *Linaria maroccana*
toothwort SEE *Lathraea clandestina*
topsoil, stripping, 44, 95–6
tormentil SEE *Potentilla erecta*
trees
as setting for meadow plants, 85
prairies, 195
strimming round, 111
woodland edges, 224

trefoil, bird's-foot SEE *Lotus corniculatus*
Trifolium (clover), 8, 11, 24, 39, 45, 60, **82**, 89, 216
T. medium (zigzag clover), 70
T. pratense (red clover), 25, 32, 41, 42, 80, 122, **122**, 133
T. repens (white clover), 25, 47, 55, 81, 121, 122
Triteleia laxa (triplet lily), 177
Triticum (wheat), 181
Trollius (globeflower)
T. 'Canary Bird', 101
T. europaeus, 133–4
trout lily, American SEE *Erythronium revolutum*
tufted grasses, 101
Tulipa (tulips), 60, 160, 170
T. 'Dillenburg', **83**
T. orphanidea, 170
T. saxatilis Bakeri Group, **170**
T. sprengeri, 170
T. sylvestris, 170, **170**
Tunbridge Wells, 42
tundra, 55
turf
buying wildflower turf, 94
creating a meadow in, 92–4, **94**
Turkey, 10–11, 65, 154
twitch grass SEE *Elytrigia repens*

U

umbellifers, 150–2
United States of America, 88, 119–20, 138, 146,152, 172, 175, 187, 188, 189, 190, 191, 198, 204, 212, 223, 224
urban planting, 45, 119, 183, **210**, 215, 218–21, **219**
Urtica dioica (nettles), 99, 106, 186, 217

V

Vacratot Botanic Garden, 65
valerian, red SEE *Centranthus ruber*
Valeriana officinalis (valerian), 70
Van Gogh, Vincent, 182
Veratrum album, 70, **71**
V. nigrum, 70
Verbascum, 221
V. speciosum, **222**
Verbena bonariensis, **188**, 190
V. hastata (blue vervain), 196, 213
V.h. f. *rosea*, 196, 213
Vermont, 130, **224**
vernal grass, sweet SEE *Anthoxanthemum odoratum*
Vernonia (ironweed), 22, 119, 206, 213
V. crinita, 206
V. fasciculata (common ironweed), 206
Veronica, 154
V. filiformis (slender speedwell), 154
Veronicastrum, 204, 213
V. virginicum (culver's root), 196, 199, 213
V.v. 'Album', 24, **193**, 213
V.v. f. *roseum*, 213
vervain, blue SEE *Verbena hastata*
vetches SEE *Astragalus*
crown SEE *Coronilla varia*
horseshoe SEE *Hippocrepis*
kidney SEE *Anthyllis vulneraria*
milk SEE *Astragalus canadensis*
tufted SEE *Vicia cracca*
vetchling SEE *Lathyrus*
Vicia (vetches), 123
V. cracca (tufted vetch), **8–9**, 22, **66**, 69, 77, 108, 123, 124, **223**
Viola (violets), 57, **10–11**, 15, 101
V. odorata (sweet violet), 145
V.o. 'Alba', **145**
V. pedata (bird foot violet), ,145
V. riviniana (common dog-violet), **144**, 145,
violet, dog's-tooth SEE *Erythronium dens-canis*
viper's bugloss SEE *Echium vulgare*
Virgil, 58
Virginia arkansana, 206
viscarias SEE *Silene coeli-rosa*
Vitex agnus-castus, 59

W

waders (birds) 35
walnut trees, **80**
wasp spider (*Argiope bruennichi*), **15**
Weald Meadows Initiative (WMI), 39, 94, 216–7
weedkillers SEE herbicides
weeds
in annual meadows, 186–7
annual weeds, 99, 183, 201
creating a permanent meadow, 99
perennial, 183, 186–7, 209, 217, 220
prairies, 198, 201, 209
weld SEE *Reseda*
wet prairie, 195, 196, 201, 203, 204, 205, 213
wheat SEE *Triticum*
'Wildflower Blanket', 94
Wiley, Keith, 152
willow SEE *Salix*
willowherb SEE *Chamerion angustifolium* and *Epilobium*
Wills, Sir John,38–9
Windsor Great Park, Surrey, 166, 171
Wisconsin, 195, 198
Wisley, Surrey, 20, 101
woodland, 11, 15, 32–4, 44, 49, 55, 89, 120, 127, 137, 141, 145, 150, 194, 224–5, 226, **227**
wormwood SEE *Artemisia*
Woudstra, Jan, 57, 60
Wright, Tom, 7, 8, 10
Wye College, Kent, 7, 10, 14, 16, 39, 124

Y

yarrow SEE *Achillea millefolium*
yew (*Taxus*)
hedging, **8**, **73**, 79,92, 123, 127
topiary, 16, 25, **26–7**, 28, **28**, 81, **81**, 84, **84**
Yorkshire fog SEE *Holcus lanatus*
Yucca whipplei subsp. *parishii* (Our Lord's candle), **194**

AUTHOR'S ACKNOWLEDGEMENTS FOR THE FIRST EDITION

The greatest debt of gratitude goes to my confederate, FERGUS GARRETT, whose interest and support have been crucial from start to finish. He has helped me with much of the research, accompanying me to Hungary (with his wife, AMANDA FERGUSON) and the USA, as well as to many of the meadows I visited in the British Isles. I am also grateful to Fergus and Amanda for their photographs, acknowledged below.

I should like to thank ERICA HUNNINGHER for meticulously editing the original text.

For photography, particular thanks go to JONATHAN BUCKLEY who has been photographing my garden since 1994 and taking pictures here at Dixter and of meadows all over the country especially for this book since I started writing it in 2000. Both he and I are grateful to the owners who allowed their gardens to be photographed.

I should like to thank the following for their help in clueing me up in their own specialized fields so as to make this a more comprehensive and authoritative work. DAWN BRICKWOOD, Weald Meadows Officer, for information on the WMI. C. COLSTON BURRELL of Native Landscape Design and Restoration, based in Virginia, for the North American wildflower scene. TOM COOPER for help on North American sources of information. NIGEL DUNNETT and JAMES HITCHMOUGH of the Department of Landscape, University of Sheffield, for sharing the results of their experiments with annuals and perennials, which are on-going. AMANDA FERGUSON of the Invertebrate Conservation Unit, Zoological Society of London, for clueing me up on insects. DAN HINKLEY of Heronswood Nursery for the North American scene and the loan of Yellow Island photos. PENELOPE HOBHOUSE for aspects of the history of meadows. GEZA KOSA, Curator at the Vácrátót Botanic Garden near Budapest, for being a most helpful host on a visit to Hungary. BEN LE BAS, English Nature National Park Officer, and JANE CHAPMAN and REBEKAH NEWMAN of the Hay Meadows Project, for the Peak District area. TONY LORD for checking idents and nomenclature. DONALD MACINTYRE of Emorsgate Seeds for many aspects of growing meadow flowers from seed. HENRY PENNER, Senior Environmental Adviser for a huge area in the south of England that includes the M25, for help on the subject of roadsides. GRAHAM ROBESON and ALAN GRAY of The Old Vicarage, East Ruston, Norfolk, for their take on the cornfield scene. JOHN SALES for demonstrating different cutting techniques and timing of meadow areas. RICHARD SCOTT of Landlife Wildflowers for aspects of the wider landscape and the loan of photos.

The internet is a means of sourcing native seeds and plants (but one always needs to check provenance) and for finding out about local initiatives. Among the many websites that have been useful for research are: www.floralocale.org (Flora locale); www.wildflower.org (Lady Bird Johnson Wildflower Center); www.landlife.org.uk (Landlife Wildflowers); www.ies.wisc.edu (Restoring the North American Ecological Landscape); and www.prairies.org.

PICTURE CREDITS

The photographs on pages 1–47 are by CAROL CASSELDEN and all those on pages 48–224 are by JONATHAN BUCKLEY, unless credited otherwise. BRIDGEMAN ART LIBRARY, LONDON & NEW YORK/Galleria degli Uffizi, Florence, Italy 59; AMANDA FERGUSON 71, 108; FERGUS GARRETT 50–51, 56, 63–68, 94, 114, 116 (below), 121, 124–5, 147, 152, 185, 191, 194, 215–6, 222–4; THE METROPOLITAN MUSEUM OF ART, Gift of John D. Rockefeller Jr., 1937 (37.80.6) Photograph © 1998 The Metropolitan Museum of Art 55.